Traditional and Modern
Natural Resource Management
in Latin America

Traditional and Modern Natural Resource Management in Latin America

Edited by Francisco J. Pichón,

Jorge E. Uquillas,

and John Frechione

UNIVERSITY OF PITTSBURGH PRESS

Published by the University of Pittsburgh Press, Pittsburgh, Pa. 15261
Copyright © 1999, University of Pittsburgh Press

Manufactured in the United States of America
Printed on acid-free paper
10 9 8 7 6 5 4 3 2 1

Library of Congress Cataloging-in-Publication Data

Traditional and modern natural resource management in Latin America
/ edited by Francisco J. Pichon, Jorge E. Uquillas, and John
Frechione.
    p. cm. —(Pitt Latin American series)
    Includes bibliographical references and index.
    ISBN 0-8229-4103-1 (acid-free paper)
    ISBN 0-8229-5703-5 (pbk. : acid-free paper)
    1. Natural resources—Latin America—Management. 2. Natural
resources—Latin America—Management—Case studies. 3. Rural
development—Latin America. 4. Rural development—Latin America—
Case studies. 5. Agricultural development projects—Latin America.
6. Agricultural development projects—Latin America—Case studies.
I. Pichon, Francisco J. (Francisco Javier) II. Uquillas, Jorge E.
(Jorge Enrique), 1946– III. Frechione, John. IV. Series.
    HC123.5 .T73 1999
    333.7'098—dc21   99-6638
                                      CIP

A CIP catalog record for this book is available from the British Library.

A previous version of Chapter 1, "Rural Poverty Alleviation and Improved
Natural Resource Management Through Participatory Technology Development in Latin America's Risk-Prone Areas," by Francisco J. Pichón and Jorge
E. Uquillas was published in the *Journal of Developing Areas* 31, no. 4 (summer
1997): 479–514, © Western Illinois University. Reprinted by permission. Another previous version of this chapter was published as "Sustainable Agriculture Through Farmer Participation: Agricultural Research and Technology
Development in Latin America's Risk-Prone Areas," Francisco J. Pichón and
Jorge E. Uquillas, in *Mediating Sustainability: Growing Policy from the Grassroots*,
eds. Jutta Blauert and Simon Zadek. West Hartford, Conn.: Kumarian Press,
1998. Copyright © Kumarian Press. Reprinted by permission.

This volume is dedicated to the memory of D. Michael Warren (1942–1998), who pioneered the multidisciplinary study of indigenous knowledge systems and who has been an inspiration for all of those involved in this field of study.

Contents

List of Acronyms

AGRTN	Agriculture Technnology Note, Central Agricultural Department, World Bank
AGRUCO	Agroecological Project of the University of Cochabamba (Bolivia)
ANAPQUI	Associación Nacional de Productores de Quinoa (Bolivia)
AOCACH	Associación de Organizaciones Campesinas Autonomas de Chimborazo (Ecuador)
APAIFO	Agriculture, Industry and Forestry Producers Association (Costa Rica)
ARCIK	African Resource Center for Indigenous Knowledge (Nigeria)
BRARCIK	Brazil Resource Center for Indigenous Knowledge
BURCIK	Burkina Faso Resource Center for Indigenous Knowledge
CABI	Commonwealth Agricultural Bureau International (England)
CAFMA	Forest Certificate Bond for Management (Costa Rica)
CARIKS	Center for Advanced Research of Indigenous Knowledge Systems (India)
CECIK	Center for Cosmovisions and Indigenous Knowledge (Ghana)
CECOAT	Center of Agricultural Cooperatives, Operación Tierra (Bolivia)
CEDESUR	Center for Development Studies—Uruguay

List of Acronyms

CENARGEN	National Center for Genetic Resources and Biotechnology (Brazil)
CEPAL	Comisión Económica para América Latina (United Nations agency)
CEPLAC	Comissão Executiva do Plano da Lavoura Cacaueira (Brazil)
CGIAR	Consultative Group for International Agricultural Research
CIAD	Center for Integrated Agricultural Development (China)
CIAL	Comités de Investigación Agropecuaria Local (Colombia)
CIAT	International Center for Tropical Agriculture (Colombia)
CIDA	Canadian International Development Agency
CIDE	Center for International Development and Environment
CIDOB	Confederación Indígena del Oriente, Chaco y Amazonia de Bolivia
CIESIN	Consortium for International Earth Science Information Networks (United States)
CIIFAD	Cornell International Institute for Food, Agriculture and Development (United States)
CIKARD	Center for Indigenous Knowledge for Agriculture and Rural Development (United States)
CIKFAB	Center for Indigenous Knowledge at Fourah Bay College (Sierra Leone)
CIKFIM	Center for Indigenous Knowledge in Farm and Infrastructure Management (Nigeria)
CIKIB	Center for Indigenous Knowledge on Indian Bioresources (India)
CIKO	Cameroon Indigenous Knowledge Organization
CIKPREM	Center for Indigenous Knowledge on Population, Resource and Environmental Management (Nigeria)

List of Acronyms

CIMMYT	International Center for Improvement of Maize and Wheat (Mexico)
CINEP	Centro de Investigación y Educación Popular (Colombia)
CIP	International Potato Center (Peru)
CIRAN	Center for International Research and Advisory Networks (Netherlands)
CODEFORSA	San Carlos Commission for Forestry Development (Costa Rica)
CODEL	Coordination in Development (United States)
COOPESANJUAN R.L.	San Juan Farmer's Cooperative (Costa Rica)
CORACA	Corporación Agropecuaria Campesina of the Federación Sindical Unica de Trabajadores Campesinos de Potosí (Bolivia)
COSEFORMA	Proyecto de Cooperación en los Sectores Forestales y Maderero (Costa Rica)
CPR	common property resources
CRSP	Collaborative Research Support Program (USAID project)
CTK	Center for Traditional Knowledge (Canada)
DBH	diameter at breast height
DESCO	Centro de Estudios y Promoción del Desarrollo (Peru)
DGF	Costa Rican Forest Service
DRI	Integrated Rural Development Program (Colombia)
El Ceibo	Central Regional Agropecuaria-Industrial de Cooperativas "El Ceibo" Limitada (Bolivia)
ELLRIK	Elliniko Resource Center for Indigenous Knowledge (Greece)
EMENA	Europe, Middle East, and North Africa region
EPA	Environmental Protection Agency
FAO	Food and Agriculture Organization (United Nations agency)

List of Acronyms

FOIN	Federación de Organizaciones Indígenas del Napo (Ecuador)
FPR	farmer participatory research
FRSE	Farming Systems Research and Extension
FSUTC-SC	Federación Sindical Unica de Trabajadores Campesinos de Santa Cruz (Bolivia)
FUNDAEC	Fundación para la Aplicación y Enseñanza de las Ciencias (Colombia)
Fundagro	Fundación para el Desarrollo Agropecuario (Ecuador)
FUNORSAL	Fundación de Organizaciones de Salinas (Ecuador)
GDP	gross domestic product
GEF	Global Environment Facility (United States)
GERCIK	Georgia Resource Center for Indigenous Knowledge (Republic of Georgia)
GHARCIK	Ghana Resource Center for Indigenous Knowledge
GIA	Grupo de Investigaciones Agrarias (Chile)
GIS	geographic information systems
GTZ	German Technical Cooperation Project
IBAMA	Institute of Environment and Natural Renewable Resources (Brazil)
ICIK	Interinstitutional Consortium for Indigenous Knowledge (United States)
ICRISAT	International Crops Research Institute for the Semi-Arid Tropics (India)
IDRC	International Development Research Center (Canada)
IDS	Institute of Development Studies (England)
IFPRI	International Food Policy Research Institute (United States)
IIED	International Institute for Environment and Development (England)

List of Acronyms

IIRR	International Institute for Rural Reconstruction (Philippines)
IITA	International Institute of Tropical Agriculture (Nigeria)
IITAP	International Institute of Theoretical and Applied Physics (United States)
IK	indigenous knowledge
ILEIA	Information Center for Low-External-Input and Sustainable Agriculture (Netherlands)
ILO	International Labor Organization (United Nations agency)
INDAP	Instituto de Desarrollo Agropecuario (Chile)
INIREB	National Institute of Biotic Resources (Mexico)
INRESC	Indigenous Resources Study Center (Ethiopia)
INRIK	Indonesian Resource Center for Indigenous Knowledge
INTSORMIL	International Sorghum and Millet Collaborative Research Support Program (United States)
IPGRI	International Plant Genetic Resources Institute (Italy)
IPRA	Investigación Participativa para la Agricultura (Colombia)
IRRI	International Rice Research Institute (Philippines)
ISAR	Senegalese Institute for Agricultural Research
ISNAR	International Service for National Agricultural Research (Netherlands)
ITP	Intermediate Technology Publications (England)
IUCN	World Conservation Union (Switzerland)
IVITA	Institute for High Altitude and Tropical Veterinary Research (Peru)
KENRIK	Kenya Resource Center for Indigenous Knowledge
LAC	Latin America and the Caribbean region
LARC	Regional Conference for Latin America (Chile)

List of Acronyms

LEAD	Leiden Ethnosystems and Development Program (Netherlands)
LKS	local knowledge system
MARCIK	Madagascar Resource Center for Indigenous Knowledge
MARECIK	Masailand Resource Center for Indigenous Knowledge (Tanzania)
NARS	National Agricultural Research Services
NAS	National Academy of Sciences
NGO	nongovernmental organization
NIRCIK	Nigerian Resource Center for Indigenous Knowledge
NISER	Nigerian Institute of Social and Economic Research
NPV	net present value
NRC	National Research Council (United States)
NRM	natural resource management
NWFP	nonwood forest products
ODA	Overseas Development Agency (England)
ODI	Overseas Development Institute (England)
OECD	Organization for Economic Cooperation and Development (France)
OPIP	Organización de Pueblos Indígenas de Pastaza (Ecuador)
PCAARD	Philippine Council for Agriculture, Forestry, and Natural Resources Research and Development
PhiRCSDIK	Philippine Resource Center for Sustainable Development and Indigenous Knowledge
PTD	participatory technology development
R&D	research and development
REPPIKA	Regional Program for the Promotion of Indigenous Knowledge in Asia (Philippines)

List of Acronyms

RIDSCA	Mexico Research, Teaching and Service Network on Indigenous Knowledge
RIVM	National Institute of Public Health and the Environment (Netherlands)
RPOs	rural people's organizations
RURCIK	Russian Resource Center for Indigenous Knowledge
SANREM	Sustainable Agriculture and Natural Resource Management Program (United States)
SARCIK	South African Resource Center for Indigenous Knowledge
SAREC	Royal Swedish Academy of Sciences
SCOPE	Scientific Committee on Problems of the Environment (United States)
SID	Society for International Development (Italy)
SLARCIK	Sri Lanka Resource Center for Indigenous Knowledge
SNV	Servicio Holandés de Cooperación al Desarrollo
SRCRSP	Small Ruminant Collaborative Research Support Program (United States)
UCASAJ	Unión de Cabildos de San Juan (Ecuador)
UCIG	Unión de Comunidades Indígenas de Guamote (Ecuador)
UMATA	municipal credit and technical assistance agencies (Colombia)
UNAPEGA	Unión Nacional de Pequeños Agricultores y Ganaderos (Bolivia)
UNCED	United Nations Conference on Environment and Development
UNDP	United Nations Development Program
UNEP	United Nations Environment Program
UNITAS	Unión Nacional de Instituciones para el Trabajo de Acción Social (Bolivia)

List of Acronyms

UOCACI	Unión de Organizaciones Campesinas de Cicalpa (Ecuador)
UPWARD	User's Perspective With Agricultural Research and Development (Philippines)
URCIK	Uruguay Resource Center for Indigenous Knowledge
USAID	United States Agency for International Development
VERSIK	Venezuelan Secretariat for Indigenous Knowledge and Sustainable Development
WCMC	World Conservation Monitoring Center (England)
WRI	World Resources Institute (United States)
WWF	World Wildlife Fund (United States)

Acknowledgments

We would like to express our sincere appreciation to William Partridge who, as director of the World Bank's Environment Unit for Latin America (LATEN), supported the research for the background paper by Pichón and Uquillas and helped convene the "Workshop on Traditional and Modern Natural Resource Management in Latin America and the Caribbean" at the World Bank on 25–26 April 1995, where the chapters in this volume were originally presented. Valuable substantive and editorial comments on the background paper were provided by Nigel Smith (University of Florida), Jutta Blauert (Institute of Development Studies, University of Sussex, England), Thomas Wiens, William Partridge, and Peter Brandriss (World Bank). Nichole M. Parker most ably undertook the formidable task of compiling and cross-checking the bibliographic references and acronyms and proofreading the manuscript. We thank the authors for their work on the chapters in this volume, and in general, for their significant contributions to studies of natural resource management systems in Latin America. Finally, it should be noted that the conclusions and opinions presented in the papers are the authors' own and should not be attributed to the World Bank, its executive board of directors, or any of its member countries.

Traditional and Modern
Natural Resource Management
in Latin America

Introduction

John Frechione

The modern development era, which began about fifty years ago, was initiated as an attempt to improve people's quality of life and standard of living worldwide. There have been occasional successes (such as some aspects of the green revolution), but with 1.4 billion people still inhabiting risk-prone environments, and still lacking the capacity to support themselves in a sustainable manner (see Pichón and Uquillas, this volume), these development efforts must be evaluated with caution in relation to alleviating poverty and resource degradation in the risk-prone areas. The scenario that most concerns development practitioners at present is one of high, and often increasing, human populations who depend on agricultural activities for their livelihoods and who inhabit fairly delimited risk-prone areas.[1]

Many of these areas are, in fact, already degraded and are likely to degrade further under currently available intensification practices. Increased agricultural production or extraction of natural resources can be achieved either by the expansion of areas under production or by more intensive management. The first option frequently results in great losses of biodiversity, accelerated soil erosion, and increased flooding, among other negative repercussions. Currently in Latin America, areas that remain for the expansion (extensification) of agriculture are marginal and particularly susceptible to environmental degradation. Many rural inhabitants are forced to choose extensification because of a lack of viable alternatives, although this option is becoming less available as agricultural frontiers close. The second option (more intensive management of agriculture and extraction) can help to alleviate pressure on such risk-prone areas, but information on strategies to achieve this goal is generally lacking.

In the development arena, much of the emphasis on natural resource management (NRM) has focused on the top-down dissemination of modern technology, based on relatively simple or homogeneous systems such as plantation forestry, monocropping with high-yielding cereals, and ranching. In contrast, the rich reservoir of empirical knowledge acquired by rural peoples is often ignored. Traditional management practices involving agroforestry, forest fallows, extraction of resources, and so on may have much to offer for developing and conserving risk-prone areas in sound ways.[2]

Natural resource management—where the interaction of humans and nature involves a broad set of strategies and technologies for a wide variety of natural resources—is presently a topic of worldwide interest. Aware of this trend and that conventional approaches have failed to deal adequately with problems of complex agroecological systems, international agricultural research institutions such as the International Potato Center (CIP) and the International Center for Tropical Agriculture (CIAT) have created NRM programs. Small-scale research and extension projects using participatory and systemic approaches have multiplied in Latin America over the last few years. Although neither the positive nor the adverse social, economic, and environmental impacts of traditional natural resource management have been sufficiently documented, it appears that traditional NRM systems may have much to offer.

Confronted with a complex of profound problems that contribute to the perplexing situation of poverty and environmental degradation in risk-prone agricultural areas, development agencies (among many others) are seeking solutions from all potential sources. The current endeavor is to seek, develop, and implement, in risk-prone areas, intensification strategies that will provide increased yields, that are sustainable (that is, they will not degrade the natural resource base in the long term), and that will contribute to the preservation of biodiversity, in itself a proved source of components for potentially sustainable strategies.

As a major role player in agricultural development efforts worldwide (and, since the early 1970s, having poverty reduction as a priority), the World Bank has a profound interest in seeking solutions to the deteriorating situation in risk-prone agricultural areas. For this reason, its regional office for Latin America and the Caribbean initiated an analysis of how and to what extent international agricultural research centers and national research programs are addressing the crucial and interrelated issues of poverty and natural resource management in rural, risk-prone areas. This led to a particular interest in documenting exceptional policies and current trends concerning traditional and local knowledge and the real or potential contribution these could make in efforts toward sustainable development coupled with increasing food production for internal consumption and the market.

To confront these concerns the World Bank convened a workshop on "Traditional and Modern Approaches to Natural Resource Management in Latin America and the Caribbean," which took place on 25–26 April 1995. The workshop brought together a diverse group of experts, charged with addressing several interrelated major themes: (1) assessing traditional resource management strategies as a prelude to any intervention by research or development programs; (2) involving farmers and other relevant practitioners in the design and implementation of research on resource management; (3) diversifying agriculture—away from excessive dependence on a monoculture commodity approach and toward production systems based on mixed cropping, especially agroforestry; (4) introducing multidisciplinary approaches to designing NRM strategies; and (5) revamping agricultural research and extension programs, with greater involvement of the private sector and nongovernmental organizations (NGOs). The results of this exercise form the chapters of the present volume.

John Frechione

Structure of the Volume

The nine chapters that constitute this volume contain a wealth of data, analyses, recommendations, and considered opinions presented by individuals who have been intimately involved over the long term in theoretical and practical work related to the themes delineated above. The chapters have been organized into four sections, although they naturally overlap in their treatment of the complex and interrelated subject matter. Broadly delineated, the sections characterize and conceptualize the problem, discuss theoretical and practical issues, present various case studies, and examine indigenous/local knowledge systems.

The first section consists of only one chapter, "Rural Poverty Alleviation and Improved Natural Resource Management Through Participatory Technology Development in Latin America's Risk-Prone Areas," by Francisco J. Pichón and Jorge E. Uquillas. The "think-piece" for the contributions that follow, this daunting overview combines macro- and microlevel information and analysis. The authors describe the broad range and interconnectedness of structural aspects related to NRM and sustainable agricultural development in Latin America and the Caribbean. They delineate a conceptual framework for approaching NRM and sustainability issues in the risk-prone areas of the region, then they move to country-level data to explore the relationships described in the conceptual framework.

The discussion is followed by a consideration of modern and traditional approaches to technology development and the relevant achievements and limitations of each type of development in contributing to poverty alleviation and improving natural resource management. In a brief analysis the authors consider the basic trends (including inadequacies) that are discernible from case studies and field experiences dealing with farmer participation in technology development and dissemination. Finally, they review the roles of the Consultative Group for International Agricultural Research (CGIAR) centers, national programs, and the World Bank.

The study by Pichón and Uquillas shows that poverty-driven environmental degradation is common throughout Latin America and the Caribbean—with the most critical areas being in northeastern Brazil, southern Mexico, and the densely settled hillside areas of the Andes, Central America, and the Caribbean. These areas pose an acute set of problems: limited potential for growth, high and often increasing population density, degradation of natural resources, and lack of skills, education, and health care due to persisting socioeconomic biases against traditional rural communities. Under such conditions, farmers are likely to fall deeper into poverty, either becoming city slum dwellers or turning to environmentally destructive cultivation of marginal lands and forest frontiers. Although in the long run out-migration may be an answer for some traditional areas, merely transferring poverty and population pressures to urban areas and the forest margins cannot be an acceptable solution to rural problems. Clearly, to achieve the overriding goal of alleviating poverty in an environmentally sustainable manner, policies will have to focus on the risk-prone environments where poverty is concentrated.

The Pichón and Uquillas study also suggests that to make poverty-reduction interventions both welcome and sustainable in traditional areas, these interventions must be based on a more participatory approach for technology development and dissemination. Such an approach should include an agenda for formulating participatory NRM strategies—using both farmer-based traditional innovations and selective external inputs—to improve and diversify rural incomes, conserve water and soils, and expand labor opportunities in agricultural areas. The authors emphasize that designing and implementing such an agenda will require the cooperation of governments, international agencies, NGOs, the private sector, and the technical and scientific communities in the region.

The second section of this volume concerns theoretical and practical issues and includes two chapters that address various aspects of the problem from different perspectives. In "Rural Development and Indigenous Resources: Toward a Geographic-Based Assessment Framework," Bruce A. Wilcox focuses on the need to bring a general systems theoretical perspective to bear on the problem of rural smallholder development. This perspective considers the biophysical, sociocultural, and economic attributes of a rural area as mutually interdependent and, in turn, as only one configuration of components of more inclusive, hierarchically structured, and interdependent systems—such that, for example, "a household economy exists within a village economy, which exists within the economy of a local region, which exists within a national economy, and so on. A similar structure can be described for biophysical systems or biological diversity generally, as well as socio-cultural systems" (Wilcox, this volume).

Such an approach is overwhelming, given the potential extent of the data and the intricacy and complexity of the interrelationships that must be taken into account. Until quite recently, the lack of tools to handle such information meant that—although it was acknowledged that these subsystems, systems, and relationships existed and were extremely important—they could only be referred to, whereas actual research tended to focus on simpler interrelationships. As Wilcox indicates, with the introduction of new tools such as geographic information systems (GIS) technology and powerful computers, more realistic and more useful data collection and analyses are now within our grasp. He proposes that indicator methodologies employing data in a GIS format are the principal means of assessing rural development needs and opportunities.

Wilcox discusses two indicator approaches as particularly useful. The first is based on the valuation of ecoregions (an example is the World Bank's Ecoregion Project for Latin America and the Caribbean). It can be used to assess the utility values related to specific land units and to determine the values/opportunities that will be lost because of environmental degradation. This GIS approach transforms complex and unwieldy data into maps, making the data accessible to a broad group of users. The second approach is based on an environmental policy indicator model. It can be used for assessing the current ecological integrity of land units and the probability of change, given present pressures; thus, it can be used to identify risk-prone rural environments. The combination of these approaches has the potential of providing a more realistic and rational basis for policy formation and development assistance in relation to smallholder development and sustainable natural resource management.

Wilcox presents us with an important theoretical framework and some practical approaches that can contribute to sustainable development in risk-prone regions. He also emphasizes that, within this framework, indigenous resources (including knowledge) must be respected and incorporated in planning and decision making. He makes a particularly cogent observation:

> enhancement of smallholder agricultural production through conventional agricultural intensification (based primarily on increased external inputs such as fuel, pesticides, and fertilizers) may not be an option for most smallholders, for both economic and ecological reasons.

Wilcox proposes that the strategies with the greatest potential in risk-prone environments are those that conform to appropriate ecosystemic principles—such as mixed cropping systems, agroforestry, and the sustainable extraction of nonwood forest products. Although he does not delve into the problem in detail, he reminds us that the implementation of such strategies depends on coordinating management efforts at the regional ecosystemic level. It is worth noting that such "management" coordination can present a significant barrier to development efforts. Janzen (1972: 80) pointed out over twenty-five years ago, "the lack of ecological planning [in agriculture] is frequently dictated by exploitation programs that are of immediate advantage to certain individuals or governments. . . . This applies on all levels: from the farmer . . . to the government."

A number of issues related to "management" problems are alluded to in the section's second chapter, "Combining Indigenous and Scientific Knowledge to Improve Agriculture and Natural Resource Management in Latin America," by Billie R. DeWalt, although these issues do not constitute the main focus of the chapter. There are innumerable stakeholders in the complex research, planning, and implementation processes related to agricultural development strategies, and the interests (or self-interests) of the stakeholders often do not coincide. DeWalt notes: "While working as part of INTSORMIL, for example, I was surprised to find almost as much hostility and back-biting among agricultural scientists of various subdisciplines as between these scientists and the social scientists."

DeWalt's perceptive chapter delineates the necessity of melding indigenous/local knowledge and scientific knowledge if progress is to be made in dealing with the problem of smallholder agricultural development in risk-prone areas of the region. He comments on the basic issues of social justice and ecological sustainability and discusses the argument by proponents of the indigenous knowledge approach that the technologies used for increased agricultural production most often fail to deal adequately with these issues. He emphasizes that both "scientific" and "indigenous" knowledge systems have positive and negative aspects and he discusses both strengths and weaknesses. With an analysis of a configuration of factors, DeWalt provides a serviceable set of terms to characterize these systems: scientific knowledge produces "immutable mobiles" (information that can be transferred without modification to any spatial or social location) whereas indigenous knowledge produces "mutable immobiles" (relatively malleable knowledge, finely tuned to the continually changing circumstances that define a particular locality).

To illustrate the strengths and limitations of indigenous knowledge systems, DeWalt examines three case studies of agricultural and natural resource management in Latin America. The first case, involving crop damage by insect pests in a small farmer agricultural system in Honduras, portrays the limitations of indigenous knowledge systems and the contribution that can be made by careful research by Western scientists. The second case, dealing with fallow management by Runa Amerindians in the tropical forest of eastern Ecuador, depicts the efficacy of indigenous knowledge for sustainable NRM and agricultural intensification in a fragile environment where science and technology have thus far had little success. It should be noted that this represents the type of intensification Wilcox proposes when conventional agricultural intensification in such risk-prone areas may not constitute a viable option. The third case, involving a search for a better method of planting by small farmers in the Mexican central highlands, indicates how inhabitants of risk-prone areas adapt models from modern farming methods to suit their own needs, thus combining indigenous and scientific knowledge. He notes: "The[se] cases . . . indicate that we should not rely solely on the findings of agricultural scientists nor on the indigenous knowledge of farmers. We should take advantage of the creativity and innovativeness of both groups." He proceeds to discuss how the "culture" of modern science and its institutional framework make it difficult to achieve this goal. Finally, he offers some suggestions and ideas on how the scientific knowledge system should change so that it can work in more mutually beneficial ways with local knowledge systems and, thus, improve agricultural and NRM systems.

This chapter should be required reading for all actors in agricultural development efforts. The increasing specialization and particularized knowledge of modern scientists create "cultures" that are frequently difficult to bridge, breed a certain degree of arrogance, and foster a lack of respect for other culture-bearers. Leaving aside for the moment the more commonly underscored problem of working in truly cross-cultural contexts, these "cultures" all function to build significant barriers to multidisciplinary work among the subcultures of Western science. DeWalt's chapter suggests that breaching even a small gap between such cultures would produce substantial payoffs for poverty reduction and sustainable natural resource management in the risk-prone areas of the region—as well as for development efforts worldwide.

In the next section, concerning case studies, three chapters provide results from experiences where farmers and other relevant practitioners have been involved in researching, designing, and implementing natural resource–related management strategies. Enhancing the role of farmers in local analysis, in setting priorities, in experimentation, and in other research and extension activities is now widely recognized as a prime professional challenge and methodological frontier. Therefore, these chapters contribute important, on-the-ground information to help meet this challenge.

In "Organizing for Change—Organizing for Modernization? Campesino Federations, Social Enterprise, and Technical Change in Andean and Amazonian Resource Management," Anthony J. Bebbington analyzes general trends in agricultural technology utilization and natural resource management by campesino federations in Bolivia and Ecuador. Combining information from a number of cases rather than focusing on only one experience, he considers how campesino organizations have become involved in such activities, and the different ways they have—or have not—sought to combine traditional with modern resource management ideas. He then discusses implications for the potentials of, and constraints on, these organizations, and how they might strengthen their role in fostering more sustainable and socially inclusive forms of agricultural development and natural resource management.

Understanding the nature and role of these organizations in general (and in agriculture in particular) is important for many reasons. First, as the role of the rural state changes and as the institutions of rural civil society are expected to assume more responsibility in various dimensions of NRM and agricultural development, these organizations have potentially important roles to play as actors both in their own right and in partnership with other institutions. Therefore, it is important to understand these federations, in order to know how best to support and work with them. Second, by addressing the logic behind their activities, development practitioners are more likely to come to an understanding of the specifics of how and why modern techniques are incorporated and how support programs might work with modern approaches in ways that are more synergistic with traditional practices.

Two of the most important factors in influencing how—and how effectively—these federations engage in agricultural and NRM development, Bebbington argues, are their origins (why and how they were formed in the first place) and the membership principles underlying the organizations. He identifies three different trajectories whereby federations come to work with technology, and he outlines three general roles they play: that of technology service deliverer, that of NRM-based social enterprise, and that of natural resource–oriented pressure group. Several cases involving these three roles are considered, and from this analysis Bebbington proposes that the latter two roles are especially important in any possible intensification of livelihoods in large parts of the Andes and Amazonia. He argues strongly that the NRM-based social enterprise role holds considerable potential, but that it requires careful and committed forms of support from donors. Finally, he suggests, the cases under consideration show that these organizations could play an integral role in interinstitutional approaches to agricultural development and rural intensification. The general tone of this chapter is not overly optimistic, however.

A more optimistic chord is struck by Jacqueline A. Ashby et al. in "Organizing Experimenting Farmers for Participation in Agricultural Research and Technology Development." Their chapter provides empirical data from action-research carried out among small farming communities in the Cauca Department of southern Colombia under the auspices of the CIAT Investigación Participativa para la Agricultura (IPRA) project. The primary goal of the IPRA project is to assess the potential for creating and institutionalizing a community-based capacity that enables farmers to be directly involved in carrying out agricultural research. The objective is not only to enhance the farmer participatory approach but also to address the practical concern of how to reduce the costs of including farmers in research. The project is an excellent example of testing the theoretical proposition (see DeWalt, also Pichón and Uquillas, this volume) that the melding of indigenous/local knowledge and scientific knowledge is required if progress is to be made in dealing with the problem of smallholder agricultural development in the risk-prone areas of the region.

Through the IPRA project, participatory research methods for adaptive technology testing are implemented by forming committees of farmers. The committees are responsible for carrying out technology testing—in conjunction with public sector agricultural research and extension agencies, as well as with intermediate organizations such as NGOs and farmer cooperatives. An integral component of the project is the development of courses and materials used for training farmers and the staff of public sector and intermediate organizations. These materials may also be employed for replicating the strategy in other areas. The farmers' committees, the Comités de Investigación Agropecuaria Local (CIALs), mobilize local leaders who take responsibility for designing and carrying out experiments with technologies previously not known in their communities.

Analysis of the results of the project indicates considerable success for a range of objectives. These objectives include institutionalizing an experimental responsibility among farmer committees; obtaining "hard data" from farmer-managed adaptive research; achieving a level of respect for the farmers, which encourages a reorientation of the priorities of bureaucratic institutions; and producing costs and coverages that compare quite favorably with some state or private sector systems. However, issues regarding the equitable distribution of project benefits among the farmers—especially women—remain concerns that need to be addressed.

The positive effects of favorable national governmental policies toward sustainable natural resource management in risk-prone areas underlie discussions in the final chapter of this section. In "Technologies for Sustainable Forest Management in the Northern Zone, Costa Rica," Carlos Reiche shows how the Costa Rican government has used fiscal incentives as a major policy tool in the forestry sector. Facing imminent shortages of forest products and services as a result of continuing forest degradation from "traditional" forestry practices, the government is promoting the institutionalization of sustained-yield management of forests through the Forest Certificate Bond for Management (CAFMA).

CAFMA compensates landowners for the opportunity costs lost in extracting only 60 percent of the standing commercial timber and leaving the remainder for future regeneration and harvesting. Participating landowners not only derive economic benefits from this policy, they also serve as examples to encourage other landowners to participate in CAFMA. So far, financial analysis indicates that forest owners are able to obtain profits while simultaneously maintaining the indirect benefits of ecosystem preservation under incentive program management plans. In this case, where traditional forestry management practices have proved destructive and nonsustainable, the adoption of relatively low-cost and simple technologies—in conjunction with management plans—offers promising possibilities, especially when tied to fiscal incentives.

The source of many real and potential barriers to socially just and environmentally sustainable development in marginal areas may be found in the government (local, regional, and national). Government administrations—for many complex reasons—often fail to implement policies (such as land reform, land titling, credit and technological assistance, basic human rights protection, infrastructural development, fiscal incentives, and so on) that favor populations in risk-prone zones. The chapter by Reiche illustrates the potential benefits to be reaped by making even only a small dent in government bureaucracies that are generally antithetic to such policies.

The three contributions in the final section of this volume assess the status of information available on traditional resource management strategies. They review how such information is being shared worldwide. They also provide examples of the roles that indigenous/local knowledge systems can and do play in confronting rural poverty alleviation and in enhancing sustainable natural resource management.

D. Michael Warren's chapter, "Indigenous Knowledge for Agricultural Development," builds on his earlier work *Using Indigenous Knowledge in Agricultural Development* (1991a). It presents an overview of events and activities that have taken place since the publication of the previous study in the area of indigenous knowledge and development, with a specific focus on its relationship to agriculture and natural resource management. Warren notes that the recommendation to include indigenous knowledge components in development planning and implementation (in order to facilitate participatory decision making, capacity building, and sustainability) was initially greeted with skepticism —if not downright antagonism—by some members of the development community. The proliferation of agricultural development and NRM projects with an indigenous/local knowledge component reveal, however, that this skepticism is being overcome, though much more must be done to meet the challenge.

The progress made to date is based on a better understanding of the forces that generate new indigenous knowledge. Warren notes that initial interest in the role of indigenous knowledge in development focused on collecting and categorizing the knowledge. At the moment, the crucial role that such knowledge plays in a number of areas—such as decision making, identification and prioritization of community problems, and discussions that result in local-level experimentation and innovation—is more clearly understood. Warren reviews the explosion in research and interest in indigenous/local knowledge systems and provides a valuable appendix on established Indigenous Knowledge Resource Centers worldwide as of March 1997. He emphasizes the need to disseminate such information through education and global networking and to provide access to the information not only to development practitioners but to the communities of discovery and to the general public as well.

In her chapter, "Local Knowledge Systems in Latin America: Current Trends and Contributions Toward Sustainable Development," Consuelo Quiroz nicely complements Warren's work by presenting an overview of the literature on development projects that document or utilize local knowledge systems (LKS) in Latin America. She begins with brief vignettes of projects and data collection efforts, categorized according to level of participation by target beneficiaries.[3] She follows this with an overview of some LKS in Latin America and comments on their actual and potential contribution to enhancing sustainable development and poverty alleviation in the region. The various approaches fall under several categories: in situ biodiversity conservation by traditional farmers, raised-field technology, land resource management by indigenous people in Amazonia, and ethnoveterinary medicine.

The tendency to generalize about, and approach simplistically, indigenous/local knowledge systems is criticized by Quiroz in her section focusing on the issue of gender and how it relates to local knowledge systems and natural resource management. As noted in a number of contributions to this volume, indigenous/local knowledge is not uniform throughout a particular group; it can vary considerably with age, gender, socioeconomic status, occupation, and so on. Noting the general tendency to overlook especially women's knowledge systems, Quiroz focuses discussion on this issue and provides some enlightening examples of women's roles and knowledge as these relate to natural resource management in Latin America. She concludes with a number of recommendations and policy implications.

An example of a current research effort in indigenous/local knowledge systems that clearly illustrates the possibilities inherent in such systems for sustainable natural resource management and development constitutes the final contribution to this volume. "Biodiversity and Agroforestry Along the Amazon Floodplain" by Nigel J. H. Smith provides an exciting inventory of potentially valuable and viable strategies from one of the world's richest environments—the seasonally inundated forests of the Amazon floodplain. Smith illustrates this unique environment, the value of which is not widely appreciated, by describing the wide range of products and services that exist therein and that are sustainably exploitable for rural and urban inhabitants alike. Because of the lack of appreciation or recognition of these resources, ranchers and farmers are currently in the process of depleting—on an unprecedented scale—the floodplain forests of the Amazon. Market forces, rather than fiscal incentives, are driving environmental change on the Amazon floodplain, but market prices do not capture the true cost of habitat destruction. Cattle ranching, in particular, is homogenizing the landscape.

Smith discusses a mosaic of local land-use systems, including the planting of tree crops, that would constitute more productive and environmentally sound approaches to harnessing the rich agricultural potential of the floodplain. He identifies some of the uses of floodplain forests by local people, underscores the value of floodplain forests as a source of new crop plants of potential widespread economic value, and highlights the role of home gardens as launching pads for agroforestry development.[4] Smith argues convincingly that the Amazon floodplain could be a significant producer of perennial cash crops, especially in agroforestry systems, and that this can be achieved only by combining local knowledge with scientific research and involving the private sector.

Sustainable Development in Risk-Prone Areas: A Feasible Paradigm

These papers point the way to some potentially useful solutions—and in specific cases, to some proven solutions—to alleviating rural poverty and ensuring sustainable natural resource management in marginal, risk-prone areas of Latin America. Although the overall tone might be characterized as guardedly optimistic, the authors in no way sidestep the considerable constraints faced by attempts to implement the types of strategies recommended. Such constraints appear omnipresent throughout all sociopolitical levels and socionatural situations that must be dealt with. Nonetheless, the proposed strategies do provide some promise that solutions are possible (though, at this time, possibly only on a small scale and for specific areas and conditions). As DeWalt notes in his chapter, the shift away from a "magic bullet" solution (which seeks a large-scale, sweeping answer to resolve a myriad of problems in one shot) to a realization that improvements of NRM systems and agriculture are "the result of many small modifications in complex systems" is a move in the right direction. This does not, however, imply that relatively macrolevel solutions should be totally abandoned in favor of a "small modification" approach. As Pichón and Uquillas stress in their chapter, answers will not be found in either/or strategies, that is, strategies focusing either on relatively macrolevel or on microlevel tactics, either on high-potential or on low-potential risk-prone areas, either on Western scientific or on indigenous/local knowledge systems. Answers will more readily be found through a considered and informed melding of appropriate strategies.

Theoretically, this appears eminently logical. But what about practical implications? Policy makers, governments, and donors such as the World Bank, with limited resources, are confronted with the fundamental question of whether to focus their attention on high-potential or low-potential risk-prone agricultural production areas. Many high-potential areas are now degraded or suffer from environmental stress, and there is doubt whether high-potential areas will have the capacity to meet future food needs in a sustainable manner. On the other hand, a large proportion of the poor live in the highly vulnerable, traditional areas where the risk of destruction of the natural resource base as a result of their survival strategies is high.

Generally, assistance providers have opted to invest in high-potential areas, especially given the acute set of problems (limited growth potential, high—and increasing—population density, degradation of natural resources, marginalization, and inadequate education and health care) faced in risk-prone rural areas. The conditions in risk-prone areas often result in farmers sinking deeper into poverty—either because they lack support and are excluded from participatory development processes or because they are displaced and join the swelling ranks of urban slum dwellers. Clearly, to achieve the overriding development goal of alleviating poverty in an environmentally sustainable manner, efforts must focus on the areas most plagued by poverty, that is, the risk-prone rural areas.

Difficult tradeoffs arise when investing in the development of risk-prone areas, at least in the short term. Public investment in traditional agricultural areas may have high opportunity costs relative to investment in areas of commercial agriculture where most agricultural growth has occurred. However, efficiency tradeoffs may be justified in terms of the greater impact on poverty and environmental degradation that can be achieved in traditional areas. Therefore, although in the short term alleviating poverty and improving natural resource management may not necessarily be linked to promoting significant agricultural growth in traditional areas, in the long run poverty alleviation is necessary for achieving sustainability in these areas.

The discussions in this volume suggest that a new paradigm be adopted, which both integrates risk-prone areas into rural growth strategies and rehabilitates degraded high-potential zones and restores their productivity to whatever extent is feasible. Experience suggests that attempts to make low-potential zones do the same things as high-potential zones are problematic. Some argue that the best policy to pursue in vulnerable, low-potential areas is to relieve population pressures by encouraging massive out-migration. It is possible that out-migration may be an answer for some traditional areas over the long-term, but relocating poverty and population pressures to forest frontiers and urban areas cannot be an acceptable solution to the problems under consideration. Failure to address problems within the traditional areas themselves will only intensify or shift the problems to cities, as the rural poor exhaust natural resources and move to join the urban poor. Although more attention must be paid to areas with fragile ecosystems and large numbers of poor people, further improvements also must be pursued in the agricultural productivity of high-potential areas to meet the needs of rapidly increasing urban populations. In a sense, the successes of the green revolution have bought time in these areas, time for adjusting the intensification approach to assure that it really is sustainable in the longer term. Moreover, policy makers need to explore the scope for assisting low-potential areas to diversify production into goods demanded by adjacent high-potential areas and thereby benefit from their growth.

To date, neither local processes of intensification nor top-down, large-scale, agricultural and technological development projects have proved adequate in alleviating poverty and resource degradation and in establishing environmentally sustainable and economically productive systems of natural resource management in the heavily populated risk-prone areas of Latin America (or the world). These environments, in general, are degraded and continue to degrade. If this pattern persists, these areas will eventually be unable to serve even as minor agricultural production zones. The problem is daunting. Among the frantic search for solutions, the melding of indigenous/local knowledge systems (specifically in relation to agriculture and natural resource management) and the Western scientific knowledge system appears to hold some promise, as the chapters in this volume affirm.

PART I

Characterizing and
Conceptualizing
the Problem

1

Rural Poverty Alleviation and Improved Natural Resource Management Through Participatory Technology Development in Latin America's Risk-Prone Areas

Francisco J. Pichón and Jorge E. Uquillas

Worldwide, almost two billion hectares of land (about 15 percent of vegetated soils) have been degraded since 1945, about three hundred million hectares of which have suffered such extreme degradation that reclamation of their original biotic functions may no longer be feasible. Although two-thirds of the world's degraded lands are located in Asia and Africa, human-induced degradation is most severe in the Andean region, Central America, and Mexico, where one-quarter of the vegetated land is degraded. In addition, about seventeen million hectares of

A number of earlier versions of this chapter were published: *Sustainable Agriculture and Poverty Reduction in Latin America's Risk-Prone Areas: Opportunities and Challenges*, Latin America and the Caribbean Technical Department, Regional Studies Program Report No. 40 (Washington, D.C.: World Bank, 1996); "Agricultural Intensification and Poverty Reduction in Latin America's Risk-Prone Areas," *Journal of Developing Areas* 31.4 (1997): 479–514; and "Sustainable Agriculture Through Farmer Participation: Agricultural Research and Technology Development in Latin America's Risk-Prone Areas," in J. Blauert and S. Zadek, eds., *Mediating Sustainability: Growing Policy from the Grassroots* (West Hartford, Conn.: Kumarian Press, 1998). The editors are grateful for permission to reproduce parts of those articles herein.

forests are cut down each year, much of it for conversion to agricultural use by farmers. Overgrazing, deforestation, and overexploitation for fuelwood have caused 70 percent of global soil degradation since 1945. To a large extent, these activities are related to poverty and the lack of opportunities for agricultural intensification. Faulty agricultural practices, which account for another 28 percent of soil degradation, may also be partly attributable to poverty (Pinstrup-Andersen, Pandya-Lorch, and Hazell 1994).

Rural poverty combined with increasing population densities and inadequate land-use systems is responsible for much of the forced exploitation of environmentally fragile lands such as forests and steep hillsides, and for the breakdown of indigenous institutions and natural resource management (NRM) systems. Poverty-induced resource degradation is a significant and persistent problem in the developing world. Wolf (1986) estimates that some 1.4 billion people (about one-quarter of the world's population) lack sufficient income or access to credit to purchase appropriate tools, materials, and technologies to practice environmentally sustainable agriculture, protect natural resources against degradation, or rehabilitate degraded resources. These people, who have lost the capacity to support themselves sustainably, live in what Chambers, Pacey, and Thrupp (1989) call complex, diverse, and risk-prone areas. Although risk-prone areas are most widespread in Asia and Sub-Saharan Africa, they are found also in northeastern Brazil and in many parts of Central America, Mexico, and the Andean subregion. It is estimated that one billion people live in risk-prone environments in Asia, three hundred million in Sub-Saharan Africa, and one hundred million in Latin America (Wolf 1986).

Population growth is a key catalyst of poverty-led environmental degradation, especially in marginal lands. Rapid population growth diminishes farm sizes and ultimately pushes people off the land to search for land and employment opportunities elsewhere. As populations concentrate in areas not yet severely degraded, they invariably speed up the degradation process. Other factors that trigger the poverty-degradation relationship may originate from loss of entitlements by the poor. Poor people may lose traditional access to land by misappropriation of common resources and by activities such as the construction of dams and the creation of wildlife preserves that take land out of use. In response, the poor may be forced to migrate to urban areas or marginal lands. Those going to marginal lands may move higher up hillsides and cut down forests for agricultural land. Large-scale migration, both within and between countries, may not only cause environmental degradation but may also result from it.

We must confront poverty if we are to prevent the poverty-degradation cycle from being perpetuated. Continuing to neglect the serious NRM problems occurring in risk-prone areas—where most of the world's poor live—will only exacerbate the problem of degradation. The stakes are so high that the World Bank and its partners in the development community cannot afford to neglect the problems of risk-prone areas. Improvement in the management of natural resources is not only linked to poverty alleviation but is also essential for achieving sustained productivity increases in such vulnerable areas. In these settings in the next few decades, agricultural intensification might be the only realistic strategy for addressing poverty and environmental problems; expansion of the agricultural frontier may no longer be feasible or profitable in many parts of the world. There is increasing evidence that agricultural productivity can be significantly increased by enhancing traditionally practiced farming systems through improved management techniques.[1] Alternative technologies and farming practices of potential profit to farmers include appropriate crop rotations; mixed farming systems with crops and livestock; agroforestry; biological pest control; disease and pest-resistant varieties of crops; balanced application, correct timing, and placement of fertilizers; selective use of pesticides; and minimum or zero tillage.

Contrary to some beliefs, agricultural development is part of the solution to rural poverty and NRM problems. For most farmers living in risk-prone areas, agriculture is not just one of a number of ways of earning a living. It is the principal livelihood and plays a key role in ensuring food security. If neglected, it will irreversibly degrade the main natural resources upon which these populations depend. Thus, agricultural development through improved technologies should contribute simultaneously to poverty reduction, to food security, and to sustainable natural resource management in risk-prone areas—three of the principal Bankwide goals.

In this chapter we hope to address the related issues as follows. The first section, "Natural Resource Management in Latin America," characterizes natural resource management in Latin America and the Caribbean region (LAC) and describes the structural factors that are likely to condition actions toward sustainable agricultural development in the region. In "A Conceptual Framework of Relationships Affecting Natural Resource Degradation and Sustainability in Areas of Traditional Agriculture," we present a conceptual framework for understanding the relationships affecting natural resource management and sustainability in risk-prone areas. The section entitled "The Conceptual Framework and Current Trends in LAC" explores the relationships we described in the conceptual framework, using available country-level data for the region. In "Agriculture, Technology, and Sustainable Livelihoods," we discuss modern and traditional approaches to technology development, as to their achievements and limitations in alleviating poverty and improving natural resource management. Then, in "Traditional Farming and Innovation: Reviewing Experiences," we present a brief analysis of the basic trends and deficiencies found in the wealth of case studies and field experiences dealing with farmer participation in technology development and dissemination. Finally, the roles of some of the primary actors in this arena are reviewed in the sections on "The Role of the CGIAR Centers and National Programs" and "The Role of the World Bank."

Natural Resource Management in Latin America

The Role of Agriculture

Natural resource management is essentially a human activity, oriented to sustain livelihoods and based on the utilization of the natural resources of a given geographic area. In the developing world, agriculture is the basis of these livelihoods because of its importance to the rural and national economies. However, in several important respects, the factors that condition actions toward sustainable agricultural development in the LAC region differ from those of other regions.[2] First, the relationship between population and natural resources is significantly less binding than elsewhere. There are still areas that are essentially unutilized, thus the option exists for development through expansion of area as well as through intensification.

Second, the region is characterized by a skewed distribution of land in rural areas, where a small proportion of people control most of the land in *latifundios*, while the majority of people have to survive on small parcels of land *(minifundios)*. Therefore, in spite of a considerable resource endowment, there is extensive rural poverty. The sustainability issue, in part, hinges on the fact that those resources that are concentrated in a few hands could be managed to generate greater output and employment, to take pressure off ecologically fragile and biodiversity-rich lands.

Third, the relationship of agriculture to poverty and employment is less pressing in LAC than in either Africa or Asia. The agricultural sector is much smaller in terms of both contribution to gross domestic product (GDP) and percentage of rural population.[3] Overall, rural population in LAC is projected to remain stable at 125 million until the year 2000, while urban population will grow from 275 to 400 million. Although these trends should enhance prospects of achieving more sustainable agriculture, there is still the problem of persistent poverty. Extreme, mass poverty continues to be found mainly in the rural sector, but the proportion in urban areas has grown from 30 percent in 1970 to 45 percent in the late 1980s. Of the group of society classified as "poor," the proportion in urban areas increased from 37 percent to 57 percent over the same period. This raises a basic question of policy for domestic food production for the urban market and its relationship to agricultural sustainability.

Fourth, in contrast to the relatively low share of GDP, agriculture also appears to be a more important source of export growth in LAC than in other regions and accounts for 33 percent of merchandize exports. This share is nearly on a par with—but more diversified than—Africa, while it is significantly higher than EMENA (12 percent), East Asia (19 percent), and South Asia (24 percent). The share of agriculture in merchandize imports is much more uniform across regions, ranging from a high of 24 percent in EMENA to a low of 16 percent in LAC. Another aspect characterizing agriculture in LAC is the significance of livestock, perennial crops, and forestry compared with annual crops, as measured by shares in gross value of production (CGIAR 1990). Fisheries fall far behind, but their share in LAC (6 percent) is still higher than in the other regions.

Although these macrolevel indicators should be recognized in assessing priorities for regional investments, the potentially significant role of agriculture as an engine of growth and poverty alleviation for the region should not be underestimated—especially when the social and economic importance of this sector in most LAC countries is taken into account. Within the LAC region, the considerable diversity among and within countries is obscured by regional aggregates.[4] Thus, although agriculture is only 10 percent of GDP for the region as a whole, it still accounts for over 25 percent of national GDP in Bolivia, Ecuador, Paraguay, Haiti, and Nicaragua. In addition, while agriculture's share in merchandise exports has fallen to 24 percent on average, it remains above 50 percent in half the LAC countries (Wiens 1994).[5] Moreover, rural population accounts for over 50 percent of total population in Paraguay, Haiti, Guatemala, Honduras, El Salvador, and Costa Rica, and the majority of the poor is rural in at least twelve LAC countries.[6] While a series of countries (including Trinidad and Tobago, Venezuela, Chile, Argentina, Brazil, and Uruguay) experienced negative rural population growth rates, Guatemala, Paraguay, Honduras, El Salvador, and Costa Rica experienced a rural population growth rate of 2 percent. Diverse agricultural performance in the face of a similarly unfavorable policy environment throughout the region (see, for example, Krueger, Schiff, and Valdes 1993) is another intriguing aspect that suggests intercountry and intracountry differences and special circumstances must be taken into account.[7]

Of utmost importance to poverty and its relationship to agriculture and natural resource management is the question of food security and nutrition. About fifty-five million people in the region suffer from food insecurity, meaning that they do not get enough food to lead a healthy, active life. Six million of these are children. Of the twenty-eight countries in the region for which information is available, five showed a negative trend, and in 1985 three countries (Ecuador, Haiti, and Antigua) could not have covered food needs, even assuming equal distribution (FAO 1991a, 1992).[8] Despite the more favorable balance for the remaining twenty-five countries, the number of undernourished people in the region was fifty-five million in 1985, and this figure is projected to reach sixty-two million by 2000 (FAO 1990, 1992).

Policy makers cannot afford to underestimate the poverty reduction impact that agricultural growth still has in these areas, especially among small-scale farmers (Wiens 1994).[9] Since a country's farmers are the custodians of many of its natural resources, agricultural development has a direct bearing on how these resources are managed. In the poorest areas of LAC and elsewhere in the developing world, the nexus of rural poverty, rapid population growth, and unsustainable farming is leading to the degradation of land, water, and forest resources—many of which are critical for sustaining the livelihood of the poor. Agricultural intensification that increases the productivity of scarce resources is crucial to relieving these pressures (Kevin and Schreiber 1992). Experience, such as that from the green revolution in parts of Asia, has shown that broad-based agricultural growth, involving small- and medium-sized farms and driven by productivity-enhancing technologies, offers a way to create productive employment and alleviate poverty on the scale required (Hazell and Ramasamy 1989). Although agricultural intensification can help reduce poverty and resource degradation problems, if inputs are mismanaged it can create new problems, such as increased pest resistance and a narrowing genetic base as large numbers of traditional crop varieties are replaced by relatively few modern varieties. When yields are raised by increasing the use of chemicals or diverting more water for irrigation, there also are increased possibilities of pollution, pesticide poisoning, and waterlogging and salinization of irrigated lands.

If agricultural intensification is to be desirable and sustainable in risk-prone areas, it must be based on more appropriate technologies and management practices. Experience suggests there can be a high degree of complementarity between the pursuit of growth, on one hand, and poverty alleviation, food security, and sustainable NRM goals, on the other—particularly when agricultural development is broadly based and involves small- and medium-sized farms (World Bank 1993b).[10] The modest government support required by smallholders in research and extension, infrastructure, and marketing can usually produce numerous jobs at a very low budget cost per job (Binswanger 1994). Thus, if progress is to be made on crucial NRM issues in the region (such as on-site erosion and desertification or off-site forest destruction from the expulsion of population to the humid tropical lowlands), a start must be made through a series of specifically targeted rural poverty-reduction and NRM initiatives. Such a strategy should include delineating an agenda for policy formulation toward a participatory NRM strategy based upon both "traditional" farmer-based innovations and "modern" science-based inputs, which improves and diversifies rural incomes, conserves water and soils, and improves the labor-absorptive capacity of agricultural areas. To design and implement such an agenda will require the joint cooperation of governments, international agencies, nongovernmental organizations (NGOs), the private sector, and the technical and scientific communities.

Social and Structural Factors

Perhaps the most significant feature of diversity in LAC's agricultural landscapes, encountered within many (although perhaps not all) countries, is rural dualism—also defined as "functional dualism" by De Janvry (1981). Such a pattern is characterized by the coexistence of a relatively "prosperous," commercially oriented agricultural sector and an impoverished, traditional (often indigenous) agriculture. The presence of such dualism in LAC's rural areas is recognized in two recent World Bank documents—one delineating regional strategies for the agricultural sector (World Bank 1993a) and the other rural poverty alleviation (Wiens 1994)—though the precise extent and forms of this dualism are not documented. The first report acknowledges that this phenomenon is "at the root of many economic and political problems . . . , threatening social harmony and posing hard choices for an agricultural strategy for the region" (World Bank 1993a:3).

Traditional farms are not just small and unable to support the increasing consumption and production demands of families; they are also generally located in densely populated marginal areas, on lands with inherently low productive capacity, characterized by poor or severely degraded soils, steep terrain, unfavorable climate, or a combination of these factors. In addition, the degradation of the natural resource base and the deterioration in the capacity for technology generation and dissemination have had particularly serious consequences in the more marginal traditional areas—mainly in northeastern Brazil, Central America, and the Andean subregion—where traditional agriculture is concentrated (World Bank 1993a).

Population growth also tends to be high in these areas and, despite often high levels of out-migration, the size of local populations seems destined to continue to increase for some time. Off-farm employment and income opportunities also are limited in risk-prone areas, primarily because such activities depend on local demand for nonfood goods and services. In turn, this demand is low because local agricultural incomes are low. Furthermore, degradation of the resource base limits opportunities for on-farm investment and undermines incomes and assets in ways that make sustained growth difficult. The outcome of this situation is a growing number of poor people in these areas with their natural resource base further degraded by their desperate quest for subsistence. Because the most vulnerable population in risk-prone areas is typically poor and of indigenous cultural background, the NRM issue is inextricably entwined with concern for rural poverty and the situation of indigenous communities (Wiens 1994).[11]

There are striking differences between the services provided by private and public agencies to commercial and traditional agriculture. Traditional farms have not benefited (as has commercial agriculture) from the development of private suppliers of technical and farm management information and counsel. Indeed, public investment in agricultural research and extension has tended to be allocated elsewhere. The more marginal areas risk being further neglected unless there is a rethinking of where public expenditure on research for technology development and dissemination is to be directed and managed. This neglect is likely to exacerbate poverty—keeping agricultural productivity low, transport costs high, and markets for inputs and outputs weak, and limiting access to institutional credit.

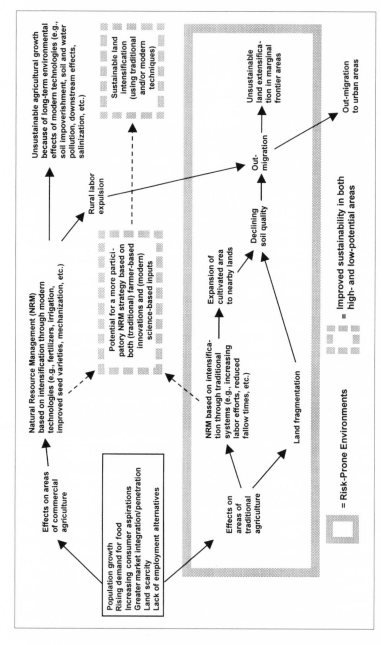

Figure 1. Integrating Traditional and Modern NRM Approaches to Foster Sustainable Agricultural Intensification and Reduce Poverty

A Conceptual Framework of Relationships Affecting Natural Resource Degradation and Sustainability in Areas of Traditional Agriculture

The issue of maintaining the sustainability of land for agricultural use is long-term and of extreme importance to LAC countries. Figure 1.1 indicates a series of possible relationships affecting natural resource degradation and sustainability in traditional agricultural areas.[12] In using the conceptual framework, one should not emphasize the inevitability or deterministic nature of the relationships, but their *possibility*.

Population growth, rising demand for food, increasing consumer aspirations, greater market integration and penetration, land scarcity, lack of alternative sources of employment, and other possible structural factors, all affect natural resource management in rural areas. If one concentrates on the effects of increased food demands on areas of traditional agriculture, it is clear that this demand can be met by an increase in the land area in agricultural use, an increase in the intensity of use of existing land, or an increase in both. The literature suggests that "families" historically tend to exhaust their economic responses prior to their demographic responses, because of the greater psychic disutility of the latter (that is, expansion of the agricultural frontier, out-migration). In other words, the family first attempts to adapt by reducing leisure, increasing labor effort, and by changing technology (to the extent that alternative technologies are known and available).[13]

The more likely response, when more land is available locally, is land extensification, however. Land extensification can be achieved either (1) by clearing more of the farmer's own land or appropriating nearby lands, without migrating away from the family plot, or (2), when such lands are no longer available and technological adaptations do not exist, by migrating away to other areas where the family might clear the trees and brush and begin cultivation anew. Depending on the area of residence, this may involve the increasing use of steep highland slopes, the clearing of brush in semiarid regions, or the cutting of trees in commonly held forested areas. Unfortunately, as populations grow and fewer unused lands with high agricultural potential remain, the land extensification process is increasingly associated with the exploitation of marginal lands (including lowland tropical forests) in the usual absence of environmental policy controls or remedial measures (Bilsborrow 1992; Bilsborrow and Geores 1992).

Provided that natural conditions such as soil quality, climate, rainfall, and so on are suitable and that the opportunity costs of land-use conversion are covered, land extensification (that is, using more land for crop and livestock production) may be a rational response to meet increasing demands for agricultural commodities (Southgate 1992). Hayami and Ruttan (1985) point out that efficient agricultural development should begin with outward shifts in agriculture's extensive margin. Investments to improve crop and livestock yields (via changes in agricultural technology, for example) are called for once opportunities for geographic expansion start to be foreclosed (Binswanger and von Braun 1991; Hochman and Zilberman 1986). In general, development of North American agriculture has been consistent with this pattern: as arable land became scarcer, growth in the agricultural sector depended more on technological innovations (Cochrane 1979).

Until recently, geographic expansion of agriculture was the major option for many, if not most, LAC countries. For example, much of the agricultural land in southern Brazil was covered with natural vegetation well into the present century. Under these conditions, yield-enhancing technologies were not critical, since expanding demands for crops and livestock could be satisfied by simply bringing more land into production (Southgate 1992). Unfortunately, current prospects for frontier expansion of agriculture are now becoming limited. The frontier is rapidly closing in areas such as southern Mexico, eastern Paraguay, northern Guatemala, Colombia, Panama, and elsewhere, where land clearing for agriculture has been the main cause of deforestation. The frontier is now primarily in Amazonia. In many other areas, virtually all soils suitable for crop and livestock production have been occupied by farmers and ranchers, and yields tend to be modest, at best, on newly cleared lands (Fearnside 1986). Moreover, since agricultural expansion is carried out largely by those displaced from older areas by poverty and land scarcity (and often by other pressures of social or political origin), the new and often most fragile lands have to be managed by those with the fewest resources to devote to their management. Under such circumstances, the need for technological innovation in agriculture is greater, but the means with which to innovate are lacking. Hence, declining productivity is compensated for primarily by bringing ever more land (usually of marginal quality) into production rather than by increasing yields per unit of area.[14] Degradation of the natural resource base in these areas is thus an inevitable consequence of poverty and inequality elsewhere, beyond the frontier.

Finally, several countries (including Brazil, Colombia, and Guatemala, among others) have developed very unequal land ownership distributions over time and have done little to change them through land reform. In recent decades, these countries imposed relatively modest taxation on their agricultural sectors and concentrated the bulk of their public sector support (such as subsidized farm credit, infrastructure investment without cost recovery, and assistance in marketing through parastatals [state companies] or statutory monopoly rights) on their large-scale commercial farming sectors. As a consequence they have fostered a dynamic, technologically sophisticated, politically articulated class of commercial farm owners. However, the fiscal burden of supporting the technologically sophisticated but economically inefficient, large-scale farming sectors has proved unsustainable both economically and environmentally (Binswanger 1994). The type of agricultural intensification taking place in these areas has generally been associated with severe changes in soil composition—reducing organic life and causing chemical contamination of soils and waters, with severe consequences for human health.[15] In addition, with the neglect of the smallholder sector, some of the excess rural population has been routed to the growing urban centers and the rest to frontier areas, usually to forests in the lowland tropics that had previously been relatively undisturbed.

In short, the overexploitation of the resource base in risk-prone areas and the expansion of subsistence agriculture to new and often marginal farming areas (extensification) have led not only to increasingly impoverished populations in both rural and urban areas (from out-migration) but also to significant environmental problems such as deforestation, soil degradation, and increased vulnerability to pest attacks, diseases, floods, and extended droughts. Many areas of traditional agriculture in the risk-prone environments of LAC such as northeastern Brazil, Central America, Mexico, and the Andean subregion are in the midst of a downward spiral of natural resource degradation and economic and social disintegration. These are some of the results of linkages between failed policies, poverty, and environmental degradation.

Of course, there are effects other than those working through the demand side, occurring directly on the supply side and involving certain time lags. For example, declining mortality means that more children survive, to enter the labor force and seek employment some twelve or fifteen years later. If farm families have plots that are insufficient to provide productive employment and if off-farm employment opportunities are not available in the vicinity, children will tend to migrate away. Second (and taking from about one generation to thirty years), having a larger number of children survive to adulthood means that existing family plots of land need to be divided among more children, which leads to the fragmentation of landholdings. Land plots eventually reach subsistence sizes that can no longer support the larger families, which again results in out-migration of either individuals or entire family units. As the out-migrants move to other rural areas, land extensification takes place again, involving the increased use of marginal lands over time, with the effects as traced above. If they migrate to urban areas, their effects on land use are indirect, resulting from increased urban consumption demands.

The Conceptual Framework and Current Trends in LAC

In this section we explore the extent to which the relationships described in the above conceptual framework are supported by available country-level data for the region.[16] In other words, we need to examine whether countries with higher rates of population growth or out-migration from rural areas have tended to experience greater degrees of land extensification (and intensification) and whether this is in turn associated with greater natural resource degradation, including deforestation.[17] When reviewing recent trends in population growth and redistribution, agricultural land area, and land-use expansion by countries in the LAC region, it is apparent that substantial differences exist.

Brazil and Mexico together account for over one-half of the total population, although their rates of population growth declined significantly between 1960 and 1990. Almost all of LAC has already experienced substantial rural-urban migration, so that the proportion of the overall urban population (70 percent) is now virtually identical to that of the developed countries. Several countries even had out-migration exceeding natural rural population growth, and hence, negative rural population growth—these are Chile, Argentina, Uruguay, Cuba, and Brazil (Bilsborrow 1992).

Changes over time in country-level statistics of the total amount of land reported as used for agricultural purposes (including animal husbandry and fallow land) show the degree of extensification of agriculture. During the last two decades or so, agricultural land increased in virtually all LAC countries, albeit to varying degrees. Bolivia and Paraguay increased arable and permanently cropped land by over 100 percent, and Nicaragua, Cuba, and Brazil by over 50 percent. Eight other countries experienced increases of between 20 and 25 percent, and the ten remaining countries somewhat less. One suspects it is the latter countries that are beginning to experience land shortages: for example, Mexico, most of Central America and the Caribbean, and perhaps the three Andean countries of Colombia, Ecuador, and Venezuela (although the latter and Brazil still have some extensive tropical lowland forest areas with agricultural potential).

Highly productive land in LAC amounts to seventy-three million hectares and is almost all under some form of production; one-third is in the temperate zone of the Southern Cone, with the remainder being in the tropics and subtropics. Sustainability issues stem from overuse and underuse. The former includes the overuse of agricultural chemicals with adverse effects on health and "downstream" effects on fisheries and recreation through water pollution. The latter includes the land's not being fully under production to contribute to economic growth and employment: such lands contribute indirectly to unsustainability and rural poverty in other regions. The Food and Agriculture Organization (FAO 1991a) indicates that nearly 10 percent of LAC's land (235 million hectares)—mainly comprising low fertility subhumid tropical savannahs—is either not utilized or is underutilized. About one-quarter of this area is currently accessible and farmed on an extensive scale. The remainder could be developed, if road and social infrastructures were forthcoming. The challenge lies in developing low-cost sustainable technologies and in providing the minimum infrastructure necessary to enable these areas to compete in domestic markets.

From the conceptual framework delineated above (see Figure 1), one would expect changes in irrigation and fertilizer use (key indicators of agricultural technology) to be positively related to growing population pressures (following Boserup's land intensification prediction). It is interesting to note that five countries in the region more than doubled their percentages of irrigated land, while six others (including the three largest in population—Brazil, Mexico, and Colombia) had increases of over one-half. Even though levels of irrigation are relatively low for the LAC region (in part because of generally adequate seasonal rainfall and greater availability of land for agricultural extensification), the median percentage of land irrigated doubled from 5.4 to 10.9 percent between 1965 and 1981—with the percentage increasing for all countries except Peru, whose percentage was already high.

Reduced productivity of irrigated agricultural lands resulting from an accumulation of salts in the soil is a major problem in all countries where a significant part of the agriculture is based on arid and semiarid lands (northeastern Brazil, Bolivia, Paraguay, central and northern Mexico, the Pacific coast of northern South America, and Patagonia). Of the total irrigated area in the region (15 million hectares), it was estimated in 1982 that salinization affected 3.3 million hectares, and the process is accelerating (FAO 1988a). Despite its adverse environmental consequences, irrigated agriculture remains crucial to overall production, particularly in countries such as Peru and Mexico.

The increase in the median for chemical fertilizer use (another common indicator of both agricultural technology and agricultural intensification) was smaller in the region, from thirty-four to forty-four kilograms per hectare. This was probably affected by the economic crisis in the region during the 1980s. The countries with over fifty, one hundred, and two hundred kilograms per hectare in 1985 numbered nine, five, and zero, respectively. The countries with the highest levels of fertilizer use were, not surprisingly, those with relatively high population densities. Countries reporting large increases were Nicaragua, Venezuela, Chile, and Ecuador—each of which more than doubled its fertilizer use in a short ten-year period.

Judging by international concern, the number one sustainability issue related to agricultural development in LAC is the expansion of cropping and livestock activities into tropical forestlands. LAC contains the world's largest remaining tropical forests, which include an enormous store of biological species. These forests play an important role in stabilizing the global climate (CEPAL 1983). During the 1970s and 1980s, it was estimated that the area of forest and woodlands in the region was reduced by 123 million hectares, about three-quarters of which were in the Amazon Basin. But in relative terms, deforestation has been faster in Central America (31 percent) and Mexico (22 percent), than in Brazil (7 percent) and the Andean countries (about 19 percent). Country-level data available on the annual estimated rate of depletion of forest cover of all types during the 1980s indicate that two countries (Costa Rica and Paraguay) reported over 4 percent annual rates of deforestation, while eight others experienced over 2 percent per year (in order, from highest to lowest: Haiti, El Salvador, Jamaica, Nicaragua, Ecuador, Honduras, Brazil, and Guatemala). Nine countries reported less than 1 percent deforestation per year. The median rate of deforestation for the countries of LAC was 1.9 percent per year—far greater than Asia and Africa (WRI 1990). This is consistent with both the large extent of forestland available in Latin America and the great increase in the amount of land converted to agriculture over the last three decades.

Another form of land degradation is soil erosion. According to FAO (1991a), soil erosion, soil salinization or alkalinization, depletion of plant nutrients and organic matter, deterioration of soil structure, and pollution have led to the loss of 10 percent of total cropland in South America and 30 percent in Central America. Land degradation is particularly serious in Central America and the Caribbean, where some twenty-five million hectares have been classed as moderately to extremely degraded. In South America, problems are acute mainly on the west coast (including the Andean subregion) and in areas where extensive deforestation has occurred (Wiens 1994). Although more detailed estimates on a country-by-country basis are not available for LAC, the phenomenon is reportedly widespread. In Central America, Leonard (1987) reports estimates for each country—for example, from five to thirty-five tons of topsoil lost per hectare per year in the highlands of Guatemala. Other country studies suggest various levels of erosion (Mexico 50 percent and Uruguay 35 percent, for example). It has been maintained that about one-half of Mexico's land is affected by moderate to advanced erosion and about one-fifth has already been lost (Goldrich and Carruthers 1992:102–03). Crosson (1983), reporting estimates of annual soil erosion based upon suspended sediments for three major Latin American river basins (Amazon, Orinoco, and Caroni), found that soil erosion was between ten and one hundred metric tons per hectare. Such data reflect a combination of the effects of agricultural practices, the extent of forest and other vegetation remaining, and density of human habitation in the watershed. Other analysts classify erosion as severe in some areas in the densely populated uplands of the tropics, although few attempts have been made to distinguish between geological and man-made causes. In addition, there is little information on the precise relationship between erosion/desertification and yield decline.[18]

It is obvious from the above review that each country in the region faces a different set of conditions that limit its strategy options in moving toward sustainable agricultural development. Land and labor availabilities are changing in important ways and the likely effects are unclear. The level of agricultural productivity in LAC has been low, whether measured by crop, meat, or milk yields. The record on crop yields is mixed. The trend in the rate of growth of cereal yields is increasing, whereas it is stagnant or declining for such staples as beans and cassava. Fertilizer use also is low relative to most other parts of the world. However, until recently, LAC (with the exception of the Caribbean, El Salvador, and Guatemala) had some remaining tropical forests, and thus high crop yields would not normally be expected. Now that the closing of the frontier is imminent, one would expect greater incentives for increasing land productivity through improved natural resource management. To investigate how resource endowments may be combined and substituted, FAO (1988a, 1991a) identified the relative importance assigned to land/labor and land/output ratios in achieving growth in agriculture. The study distinguishes four patterns of change in resource use in the region:

(1) *labor use* where growth was achieved by greater expansion in output per worker than in output per hectare, that is, in Argentina, Bolivia, Brazil, Chile, Cuba, Dominican Republic, Paraguay, Surinam, and Uruguay;

(2) *land use* where more of the growth is explained by yield increase than by output per worker, that is, in Ecuador, El Salvador, Guatemala, Honduras, Mexico, and Nicaragua;

(3) *neutral factor use* where changed labor and land productivity contribute equally to growth, that is, in Colombia, Costa Rica, Haiti, Panama, and Venezuela;

(4) *recessional use of factors* where productivity of both land and labor declined, that is, in Guyana, Jamaica, Peru, and Trinidad and Tobago.

Strategies for sustainable agricultural development vary widely according to the mix of land resource endowments available to a particular country, its human resources (urban and rural), its trading options, and the relative weight placed on development objectives. This latter area may be illustrated by extensive versus intensive development; the relative importance of agriculture in generating employment throughout the economy (rather than exclusively in agriculture), and the related question of population pressure; export orientation versus food self-sufficiency; food security or import substitution; weight assigned to strengthening agricultural linkages within the economy (industrial development as a function of agriculture); and policy discrimination against agriculture. These elements condition—and will continue to condition—the choice of agroecological zones and farming systems to be employed, that is, the agricultural strategy (FAO 1991a).

Agriculture, Technology, and Sustainable Livelihoods

During the past three decades, there has been increasing concern about the pervasive nature of poverty in the developing world (World Bank 1993b). This increase in concern coincided with the drama of major biological breakthroughs in food production associated with the "green revolution," which in turn sparked a debate on whether there was a causal relationship between the technologies of the green revolution and the incidence of rural poverty (see, for example, Mellor and Desai 1985; Rhoades 1988; Thrupp 1989). Questions arose from the realization that resource-poor farmers stand to gain little from the processes of development and transfer of technology characteristic of the green revolution, namely, the breeding of early maturing, fertilizer-responsive varieties and their diffusion into environments enhanced by irrigation and agrochemicals. Other questions have arisen based on the realization that official research and extension efforts have benefited mainly the powerful commercial farming interests (Bebbington et al. 1993b).

Critics of the green revolution have pointed out that the new technologies were not scale-neutral, that the larger and better-endowed farm areas and farmers gained most, that others often lost, and that income disparities were often accentuated. Although subsequent studies showed increasing spread of high-yielding varieties and benefits among smaller farmers in green revolution areas where they had access to irrigation, disparities remained. More significant, perhaps, was the fact that complex, diverse, and risk-prone agriculture remained poorly served by the transfer-of-technology approach. In these areas, the top-down transfer-of-technology approach has favored scientific, "modern" knowledge and in turn has neglected local participation and traditional knowledge.

Modern Technologies and Natural Resource Management: Achievements and Limitations

The introduction of high-input agriculture under the green revolution initiative channeled scarce investment resources into capital-intensive agriculture in some areas. In turn, these areas became dependent on imported machinery, equipment, hybrid seeds, chemical inputs such as fertilizers and pesticides, and irrigation. The new capital-intensive and highly market-oriented technologies worked well where ecological conditions were relatively uniform (in irrigated areas, for example) and where delivery, extension, marketing, and transport services were efficient.

Increased needs for agricultural products and the development of modern high-yielding varieties of several food crops (such as maize, wheat, rice, and other commercial crops) made the green revolution technologies attractive in many areas of high potential, and especially in Asia. This resulted in a substantial contribution to the global increase in food production. The Consultative Group for International Agricultural Research (CGIAR) estimates that, during the last two decades, agricultural research and development (R&D) have increased food production by an amount sufficient to feed one billion additional people. According to Alexandratos (1988), from 1961 to 1985 the global average yields of these major food crops rose impressively: by 41 percent for rice, 45 percent for maize, and 70 percent for wheat. New wheat and rice varieties now cover approximately twenty-four and forty-five million hectares, respectively, or 50 percent and almost 60 percent of the total planted area of each crop (CGIAR 1985). In addition, the use of hybrid maize varieties has rapidly expanded in several regions and by 1986 covered about 25 percent of the area planted with maize in Africa (CIMMYT 1988).[19]

The use of modern inputs has permitted a great increase in land-use intensity. Modern varieties are essentially high-response varieties, bred to respond to high doses of chemical fertilizers. If they are sown under conditions of high nutrient and water supply and adequate pest control, modern varieties and hybrids can and have been high-yielding. According to FAO, the enormous increase in fertilizer use was the most significant factor in raising agricultural productivity—in combination with other inputs such as access to irrigation and relevant information (Alexandratos 1988). However, when these conditions cannot be guaranteed (such as in risk-prone environments), risks of yield losses may be higher than with local varieties. Moreover, according to Dover and Talbot (1987), perhaps as much as 80 percent of agricultural land today is farmed without, or with very little, chemicals, machinery, or improved seeds.

The discussion by Reijntjes, Haverkort, and Waters-Bayer (1992) of modern agricultural research and its bias toward high-potential areas, export crops, and better-off farmers suggests several other criticisms.

(1) The emphasis has been on maximizing production of particular commodities and not total farm production. This has hindered the study and improvement of positive interactions between different plants and animals that, in addition to providing farmers' liveli-

hoods, can contribute to the continuity and stability of farming. In addition, together with other factors, the promotion of modern varieties has led to the disappearance of many indigenous varieties. This may spell disaster for farmers who have to produce their crops with low external inputs under highly variable and risk-prone conditions.

(2) The long-term effects on soil fertility, the regenerative capacity of natural vegetation and fauna, human health, and so on have not usually been given sufficient consideration because of the focus on single crops. The recent stagnation in production increase has raised serious doubts about whether long-term productivity of modern technologies is secure. Dover and Talbot (1987), for example, as well as many others have noted that the tendency toward excessive use of modern inputs is leading to soil impoverishment in numerous green revolution areas.

(3) Until recently, conventional agricultural research has paid little attention to important questions about women's influence on decision making and labor allocation when designing new technologies and extension systems.

(4) The conventional top-down approach to technology development within agricultural research institutions—that is, that the formal system is the ultimate source of innovations—has given scientists little opportunity to become well acquainted with traditional knowledge concerning ecologically oriented husbandry and local alternatives to purchased inputs. The production conditions of experimental stations seldom resemble those of farmers. Consequently, technology tested at such stations has not worked under farmers' conditions, while good qualities of local varieties (which are adapted to local conditions) are not recognized under station conditions. The "products" delivered for extension, therefore, have tended to be incomplete and designed without sufficient regard for household issues such as risk spreading, labor allocation among other existing crops, affordability of modern inputs, and other crucial aspects of the socioeconomic context.

(5) Farmers who are given access to credit may be required to engage in high capital investments and production methods that demand high levels of external inputs, which must be maintained or increased. However, when purchased inputs are subsidized by the government or a development project, their use is feasible only for a limited time. As soon as subsidies are removed and farmers are forced to abandon the inputs, it is unlikely that they will be able to adjust other aspects

of their farming systems (such as reduced diversity of crop and live-stock species or increased nutritional dependence on crops like maize, which require high fertilizer inputs) and revert back to original conditions.

The implications of modern technologies for income distribution in rural communities also have been the subject of much debate. Since modern technologies save land by permitting more intensive use of labor and external inputs, they might be expected to contribute to a more favorable income distribution among farm households. However, the new biological and chemical technologies are often criticized for benefiting landlords at the expense of tenants and laborers in high-potential areas. Land rents, for example, often increase where modern crop varieties are introduced. Empirical evidence (Hayami and Ruttan 1987) does, however, show that the adoption of modern varieties combined with irrigation and fertilizer has generally resulted in increases in labor demand, even in areas where accompanied by concurrent progress in mechanization. Despite this evidence, however, many analysts have documented growing inequalities emerging in many green revolution areas. The extent to which these inequalities can be attributed to the new technology itself or to insufficient progress in its development and diffusion is, however, unclear.[20]

Challenging critics of modern agricultural technology, Hayami and Ruttan (1987) have posed the question of whether the development of green revolution technologies should have been withheld because of their possible adverse effect on income distribution. In the absence of the new technology, the authors argue, many developing countries would have moved several steps closer to the Ricardian trap of economic stagnation and even greater stress over the distribution of income. As population growth presses against limited land resources under existing technology, the cultivation frontier is forced onto more marginal lands. As a result, greater amounts of labor must be applied per unit of cultivated land, with the result that the cost of food production increases and food prices rise.

Another reason often cited for encouraging the development and diffusion of new biological and chemical technologies—even in societies characterized by inequitable distribution of wealth and power—is that the new income streams generated by technical change represent a powerful source of demand for institutional reform. The gains from the new technology can be fully realized only if land tenure, water management, and credit institutions perform effectively. Markets for the inputs required by the new technology (seeds, fertilizer, pesticides, and so on) must perform efficiently. Hayami and Ruttan (1987) stress that in a society in which technology is static and marketable surpluses are not increasing, there are few gains either to producers or consumers from the reform of market institutions. However, when rapid growth of production and productivity becomes possible, the gains become larger and the incentives that act to induce institutional reforms become more powerful. Similarly, unless the potential gains from land tenure and other institutional reforms are enhanced by technological change, it will be difficult to generate the effort needed to bring about reforms.

The crucial question is whether advances in indigenous technology would be sufficient to sustain rising levels of per capita income and consumption—or, at least, what role should indigenous technology and management systems play in helping attain such goals. The impact of population growth on inducing indigenous improvements in agricultural technology in contemporary farming societies has been documented by an increasing literature based on Boserup's (1965, 1981) thesis. Her insistence on the importance of population growth in inducing the development of intensive systems of agricultural production is an important correction to the view that agricultural technology in traditional farming communities was essentially static. However, with some exceptions, such advances have rarely been rapid enough to do more than slow the rate of decline in labor productivity and natural resource degradation.

Building on Traditional Knowledge

Prior to the institutionalization of research in the last century, technology had evolved over millennia through natural selection and farmer selection of crop varieties and the evolution of materials and methods (Farrington and Martin 1988). Pingali and Binswanger (1987) explain how societies have been able to achieve agricultural growth resulting from innovations that are farmer-based rather than inputs that are science-based. Adaptation of farming systems, the invention of new technologies, and related investments were—and continue to be—generated by farmers themselves. Empirical evidence from several parts of the world shows that most agricultural technologies in use in the world today were developed by farmers, not by scientists at international and national research institutes (Biggs 1989; Chambers, Pacey, and Thrupp 1989; Roling 1988). Although these solutions have often been sufficient to cope with the relatively low rates of growth in demand from increasing populations, farmers' own methods of technology discovery and natural resource–based investment have not been sufficient to accommodate modern rates of growth in demand in developing countries (Binswanger and Ruttan 1978; Hayami and Ruttan 1985). These countries have been able to achieve high rates of agricultural growth only by incorporating science-based technologies such as fertilizers, pesticides, high-yielding varieties, and so on, with farmer-generated technologies (Pingali and Binswanger 1987).[21]

In many areas, especially where farmers depend mainly on local resources, modern technologies may not, however, be the first or only option to improve agriculture. Rather, improving the insight of farmers and development agents into the ecological principles behind farming and adding to their knowledge of the available technical options are important steps in the process of strengthening farmers' capacity to develop and manage technology for sustainable development. This also implies that the solutions to farmers' problems will be as diverse, complex, and site-specific as their farming systems, but that the principles involved in finding the solutions will have wider validity.

Long-term concern for the sustainability of the natural resource base is an important traditional management objective that has often been institutionalized in local regulations and cultural norms (Warren 1991a). However, where poverty has become so extreme that day-to-day survival can be the only goal, traditions of natural resource conservation may be lost (Rhoades 1988).[22] The issue is that there is no margin for reconciliation with ecological rules at the present levels of poverty found in risk-prone areas. Resource degradation in these areas may be deepening, as new technologies to intensify land use in sustainable ways—appropriate to the conditions of such risk-prone environments—have not been developed or are not known to farmers. These farmers are forced to exploit their resource base beyond its carrying capacity. In particular, this seems the case where the farmers have been deprived of access to better-quality land reserved for commercial cropping or ranching under traditional (and misguided) agricultural policies (Binswanger 1994).

The crucial issue in improving agriculture in traditional areas is not, therefore, whether one type of technology should replace the other, but how, in terms of methods and institutions, the most relevant aspects of each can be brought to bear on the NRM issues of a particular area. Agricultural sustainability in risk-prone areas requires an understanding of the diverse and complex environments in which resource-poor farmers operate, so that developments in technology can be tailored to suit their circumstances and built, where possible, on farmers' indigenous technical knowledge. Innovations combining both modern and traditional approaches to natural resource management (such as combining chemical and organic fertilizers, appropriate forms of green manuring, or integrating new crops) can open new possibilities for farmers. Spreading knowledge about these technical options and combining the forces of farmers, fieldworkers, and scientists in discovering the opportunities and limitations of these options definitely must play a role in any sustainable agricultural development strategy for the countries involved.

It would be naive to argue that farmers living in areas of traditional agriculture are not interested in incorporating modern agricultural technologies into their farming systems. To ensure the continuity of their livelihoods, farmers must be able to adjust to change driven by increasing demands—population growth, greater market integration, desire for more consumer goods, and so on. The capacity to adapt to changing conditions ultimately determines the sustainability of agriculture. Vital to such adaptability at the farm level is the capacity to manage farm development: to choose appropriate combinations of genetic resources and inputs, to develop new technologies, and to fit innovations into the farming system to raise output in subsistence food crops and commercial crop production. In this context, resource-enhancing techniques are particularly important, as they can be used not only to rehabilitate degraded land but also to create new opportunities as new needs arise.

Formal institutions of agricultural research and extension should, therefore, not be the sole agents of innovation and dissemination of new technologies. Farmers are continuously developing technology on their own but, without outside assistance, cannot advance as far or as quickly as would otherwise be possible. Scientists also have been developing technology on their own, but their impact would be greater and more beneficial if they worked more closely with farmers. Farming strategies to promote agricultural growth, alleviate poverty, and encourage the use of technologies that increase productivity while conserving soil, water, and other natural resources also can be sought from (and found in) various sources: agroecological science, indigenous knowledge and farming practices, new directions in conventional agricultural science (such as systems approach, biotechnology, and so on), and the many practical experiences of experimental farmers and fieldworkers.[23]

Traditional Farming and Innovation: Reviewing Experiences

There are no technical blueprints for sustainable natural resource management under the above conditions, but farming techniques that are most likely to be applicable in the various contexts are those that involve careful conservation of soil and water; use of complementary or symbiotic genetic resources (intercropping, integrating trees and animals, and so on); taking advantage of nitrogen fixation; and complementary and efficient use of external nutrient inputs (natural or artificial). Many traditional practices, not all of which are yet known to formal science, represent at least the seeds of promising new technologies based on composting, green manuring, mulching, multiple cropping, contour farming with bunds or hedges, water and nutrient harvesting, and ways of controlling pests. If these indigenous practices are well understood in formal scientific terms, it may be possible to improve them (for example, by the careful use of external inputs). In addition, many indigenous (sometimes unconventional) crop and animal species and local varieties and breeds have great potential for new technologies in risk-prone areas.

A growing number of publications focus on indigenous knowledge systems and the farming systems based on them (Biggs and Clay 1981; Brokensha, Warren, and Werner 1980; Chambers, Pacey, and Thrupp 1989; Marten 1986; Pretty 1995; Reijntjes, Haverkort, and Waters-Bayer 1992; Rhoades 1984; Richards 1985; Scoones and Thompson 1994; Warren 1991a; Warren and Cashman 1989; Wilken 1987).[24] A common theme running throughout most of this work is that traditional farming systems are in constant change, continually trying to adapt to the new conditions imposed by population changes, greater aspirations, market integration, and so on. However, these adaptations have not always been adequate, and entire cultures have disintegrated as a result.[25] Many indigenous practices that in the past sustained human populations for centuries have become obsolete as conditions changed. For example, several forms of shifting cultivation have proved nonsustainable under increased population pressure and, as a result, cannot be maintained without considerable environmental damage.

Not all traditional systems have reached a point of causing ecological damage, and those in the process of decline can benefit greatly from modern technology interventions to increase stability and productivity of the farm system while conserving the natural resource base. There are now hundreds of studies that have recorded the importance of traditional knowledge systems in natural resource management in many countries. Case studies of successful experiences include experimentation and innovation in food and tree crops, irrigation and other water harvesting techniques, gardening, seed distribution, field and seed preparation, fertilization, livestock nutrition, rodent and weed control, food storage, food processing, and market products and outlet, among many others.

The process of combining local farmers' knowledge and skills with those of external agents to develop or adapt appropriate farming techniques has been termed participatory technology development (PTD). In the PTD process, farmers work together with professionals such as researchers and extensionists from outside their community in identifying, generating, testing, and applying new technologies. PTD seeks to strengthen the existing experimental capacity of farmers and encourage continuation of the innovation process under local control (Haverkort et al. 1988).

Numerous case studies illustrate the range of initiatives in PTD. From a review of conceptual approaches to PTD and recent case studies, we find many instances of successful participation in problem identification, including substantial reorientation of initial objectives defined by researchers. Of particular importance for technology development is the capacity of farmers to understand the local biophysical and cultural environment and to predict and explain the outcome of experiments under local conditions. In a substantial number of cases farmers' evaluation of technology has provided researchers with new insights, and farmer-to-farmer dissemination has been successful. Where researchers already have good knowledge of a technology, a common cost-benefit technique is to offer several technology options having a bearing on the problem or opportunity at hand, and to leave it to the farmers to experiment in an ad hoc fashion.

As they become familiar with new technology, farmers also are likely to change other components of their farming system to exploit the advantages the new technology offers. Such changes can be complex and variable over time and space, so that researchers have little prospect of predicting outcomes on the basis of their own trials. Observation by researchers of the evaluation criteria used by farmers can then be integrated into the next round of technology development for release to farmers (Farrington and Martin 1988). The search for new participatory methods has led also to efforts to meet both researchers' and farmers' requirements in a single set of trials, usually through interfarm instead of intrafarm replication. These trials have been particularly useful in accelerating the release of new genetic material (Maurya, Bottrall, and Farrington 1988), though in other cases they have incurred both a substantial cost and a high risk of uninformative failure, prompting a move back to on-station trials.

Most of the material reviewed has implied participation at the individual farmer level, but other important opportunities for participation should not be neglected. Important divergences of obligations, rights, technical knowledge, and therefore acceptability of technologies have been found within farm households, which implies a need to involve women in technology development. Other experiences have shown the community to be the more appropriate level of participation, especially when dealing with technologies concerning the exploitation of common resources. Other technologies, such as innovations in animal-drawn equipment, have traditionally developed through interaction between farmers and artisans, which implies the need to build on these channels of local knowledge.

Regarding the types of technology being developed, there are many more field experiences reported in the selection of genetic material than in any other application. Important but isolated examples have been recorded in the management of soil, water, forest resources, crops, and storage facilities. Examples from crop protection, fertilization, farming equipment, and livestock research are less numerous. Perhaps the focus on genetic material highlights the area of greatest complementarity between researchers and farmers. The farmers have a vast range of material on which to draw and have developed breeding methods exceeding, both in scope and speed, those available to the researchers. It is remarkable that more than 80 percent of the crops cultivated in developing countries are planted with seeds saved from the preceding season and from the informal farmer-to-farmer seed systems (such as the informal potato seed systems in Peru).[26] Moreover, for self-pollinated crops such as rice and wheat and for crops grown primarily for subsistence such as dry beans, millet, and cassava, the proportion of farmer-saved seed is generally higher. These informal systems are strong even in countries that have relatively advanced seed industries (Jaffee and Srivastava 1992).

We also observe a tendency throughout the literature to describe the intention and rationale behind farmer involvement in technology development, but to give only a brief summary of the procedures and problems, thus making it difficult to assess the success of participation in practical terms. Much closer attention needs to be given to the role of extension, for example, once researchers and farmers have been drawn closer together in a participatory approach. The role of local organizations and nonindigenous NGOs in articulating client demand for—and mediating participatory inputs into—agricultural research has received little attention yet appears to offer considerable potential.[27] There are numerous case studies of projects using innovative methods at the outset that, to date, have not produced substantive evaluation of their experiences. Perhaps more important, the time- and cost-related effectiveness of participatory methods is poorly documented.[28] In addition, although field experiences all recognize that participatory approaches can lead to greater cost effectiveness not only in problem-focused but also in commodity and factor research, precisely how the results of the participatory methods influence the agenda for research in these areas is rarely illustrated with empirical evidence.

Concerning the institutional framework, we note that practically all PTD experiences have been undertaken outside national agricultural research programs. Numerous examples have had a continuing institutional base through the CGIAR centers, NGOs, and universities. However, many more have emerged from specific research projects of limited duration with no apparent commitment to their eventual incorporation into any institutional framework. It is clear that many more cases are needed of instances where the incorporation of PTD into national agricultural research programs has been attempted.

As a result of decreasing levels of both government and international donor funding, we also observe that NGOs are moving to fill vacuums created by the decline of extension services and are assuming greater responsibility in identifying and distributing required inputs (such as suitable seed) and providing technical support services in risk-prone areas (Farrington and Bebbington 1994). Some funding agencies such as the Ford Foundation have supported collaboration between government and nongovernment entities, utilizing the latter's capacity as "brokers" between farmers and research services. The capabilities that NGOs bring to bear in doing this are derived from their close knowledge of the needs and opportunities of the rural poor in relation to agricultural change. This holds true not only for crop or animal technology but also in the wider context of innovation located in systems that (spatially) go beyond the farm boundary to embrace the use of off-farm biomass and that (sequentially) go beyond farming systems into processing and marketing.

Although in most cases NGOs have occupied service delivery roles (effectively assuming activities and interventions that conventionally lie in the domain of government), we find some experiences of NGO involvement in development and dissemination of technologies and improved management systems among the rural poor (Farrington and Bebbington 1994). Thus, information is required concerning the long-term incorporation of PTD methods into institutions such as NGOs and universities, outside national agricultural research programs. As inevitably occurs with any new approach like PTD, we note that methods are being developed in a piecemeal fashion, although proposals for more systematic participation do exist and merit empirical testing. More careful documentation is needed on the types of PTD methods most easily incorporated; the features of institutions likely to facilitate their incorporation; and the types of institutional change that would be most conducive to fuller incorporation of technology generation methods.

Democratization of political processes and trends toward decentralization offer hope of greater accountability in agricultural development strategies (see below). Existing centralist and top-down approaches to technology development and dissemination are challenged throughout the work reviewed—not merely in technology design but in support facilities such as local gene banks and seed multiplication and in legislative provisions such as government certification and release of new varieties (Maurya, Bottrall, and Farrington 1988). It is evident that, as the variability of agroecological conditions under which farmers in risk-prone areas operate becomes better recognized, pressure will increase for a paradigm shift away from central control and toward local control and a blend of locally and centrally available support facilities. The wide range of biomass on which farmers in these areas must draw for supporting their agricultural livelihood and the diverse criteria by which they will assess the increased production and welfare benefits offered by technological change are of critical importance to the technology development process. Scarce public and private resources simply will not be sufficient for scientists at national and international research centers to undertake this task, even if they could grasp the opportunities and constraints in all their diversity. This remains one of the most compelling reasons for promoting farmer participation in the development of technology.

The Role of the CGIAR Centers and National Programs

Through their training of national scientists, their international networking of research programs, their publications, and their prestige, the CGIAR international agricultural research centers propagate and sustain the dominant concepts, values, and methods of agricultural science. As a source of development and dissemination of green revolution high-yielding varieties of basic food crops such as maize, wheat, and rice, the CGIAR system has been the subject of considerable criticism for concealing weaknesses and distortions of the new technologies in relation to how they address the crucial issues of poverty and sustainability (Chambers and Pretty 1994; Pretty 1995). Although the Technical Advisory Committee (TAC) of CGIAR has noted that all the commodities research by the CGIAR centers is relevant for the low-income group (because these commodities comprise the basic food for this group [CGIAR 1985]), critics have argued that choice of commodity has not in itself ensured benefits to the poor. There are several pertinent questions not answered simply by the choice of crops: Has research led to adoption? who adopts? and where? who gains? and who loses (women, men, the better-off, the poor, producers, consumers)? Critics argue that, notwithstanding the rhetoric, the centralized technology transfer approach practiced and propagated by CGIAR centers has tended to be insensitive to local contexts, traditional agricultural practices, and the needs and interests of their poorer clients.

The shortcomings of commodity-based research have been acknowledged by the new CGIAR focus on ecoregions, which recognizes that response to technology is conditioned not only by biological and climatic factors but also by socioeconomic and political characteristics that define behavior. The approach is expected to operate on a regional basis, to focus on important degradation problems, to combine NRM and production objectives, to employ a multidisciplinary approach, and to ensure global coherence and flexible funding mechanisms. Proponents of the ecoregion approach also recognize that it will require an unprecedented level of collaboration, negotiation, and coordination between all CGIAR centers and between the centers and the national agricultural research programs.

Although farmer participation in technology development and dissemination is not listed among the organizational principles for the ecoregion approach, groups of scientists within some CGIAR centers have been actively involved in identifying the socioeconomic aspects of farmers' problems to complement biological research efforts. Many CGIAR social scientists have already been conducting successful participatory research in partnership with other organizations and groups such as the International Maize and Wheat Improvement Center (CIMMYT) (Collinson 1985), the International Center for Tropical Agriculture (CIAT) (Ashby 1987; Ashby et al. 1990), the International Potato Center (CIP) (Rhoades 1987; Rhoades and Booth 1982), the International Service for National Agricultural Research (ISNAR) (Biggs 1988), and the International Rice Research Institute (IRRI) (Fujisaka 1991).

Experience with Farming Systems Research and Extension (FSRE) projects has provided additional evidence for the importance of understanding and building upon farmers' knowledge and decision-making processes (Rajasekaran and Warren 1991). Other international development agencies also are exploring participatory ways in which local knowledge can facilitate project efforts. Social scientists working with the U.S. Agency for International Development's (USAID) Small Ruminant Collaborative Research Support Program (SRCRSP) have been successful in combining both folk and scientific approaches to animal health problems (McCorkle 1989). Canada's International Development Research Center (IDRC) (Matlon et al. 1984), the London-based Overseas Development Institute (ODI) (Farrington and Martin 1988), and the International Institute for Environment and Development (IIED) (Scoones and Thompson 1994) have been particularly supportive of projects that incorporate traditional NRM systems. The Information Centre for Low-External-Input and Sustainable Agriculture (ILEIA) in the Netherlands and the Center for Indigenous Knowledge for Agriculture and Rural Development (CIKARD) in the United States have played important roles in establishing global networks of development professionals involved in projects that build upon traditional knowledge through participation in technology development and dissemination.

It is evident from the above review that enhancing the role of farmers in local analysis, setting priorities, experimentation, and other research and extension activities is now widely recognized as a prime professional challenge and methodological frontier. At a time when donors are looking carefully at new ways of making research more efficient (see below), national and international agricultural research institutions are being challenged to give increased attention to natural resources management and the environment as well as to meeting production needs. This will involve adjusting the research effort to the resources available, involving a wider range of actors in meeting concerns of sustainability and poverty alleviation, and making research more accountable to its clients.

In a 1993 report, the TAC/CGIAR noted that a second green revolution will be needed to double food supplies in the next thirty or forty years. To achieve this, the challenge is "to develop food production systems on *existing* farm land that will double present output levels without degrading the natural resource base on which sustained production depends, without negative effects on environmental quality, and with positive effects on the welfare of rural and urban communities" (TAC/ CGIAR 1993:2; emphasis added). As NRM and environmental concerns grow, the public research establishment is led to collaborate with universities, NGOs, and international agencies in "consortia" that bring together the strategic research and grassroots contacts needed to address sustainability issues correctly.

The Role of the World Bank

The inability of governments to increase employment opportunities and alleviate poverty in the traditional densely settled rural areas (particularly highland areas in the Andean, Central American, and Caribbean countries), combined with default on agrarian reform, has resulted in implicit or explicit sanctioning of colonization in forested areas. This phenomenon and its underlying causes have been sorely neglected by the governments concerned, and the World Bank (historically more concerned with raising agricultural productivity) has not done much to overcome this neglect (Wiens 1994). Even though the issue of sustainability provides a powerful argument in favor of agrarian reform (a critical instrument for the reduction of rural poverty [CEPAL 1988]), there is little indication that—despite reform laws, new agencies, and international pressure—the social institutions maintaining the status quo are in the process of any fundamental change. Under such conditions, there can be little doubt that deepening poverty in highland areas (and the associated acceleration of rural-to-rural migration directed toward the forest frontier) poses the principal threat to sustainable agricultural development in the region.

The opposition to land reform in many countries (Colombia, Brazil, and Guatemala, for example) resulted in the adoption of integrated rural development directed at smallholders, as the next best strategy. Ever since the World Bank made rural poverty a priority in the early 1970s, the Bank has advocated smallholder development—through integrated rural development programs and with subsector programs for irrigation, research, and extension as important strategic elements. The projects consisted typically of integrated packages of support, to smallholder agricultural development, for a specific area or region, including interventions in agricultural research and extension, marketing, input supply, credit, rural roads, water supply, electricity infrastructure, and small-scale irrigation. Often the projects also included social infrastructure such as primary schools and health centers. Some of these projects were called area development projects.

The smallholder strategy had far from universal support, however, even among Bank staff (Binswanger 1994). It was undermined especially in Bank support for agricultural credit. Although these projects consistently tried to limit subsidies and to direct more credit to the smallholder sectors, good intentions affirmed in loan documents and covenants were often undermined in practice. According to Binswanger (1994), this failure to prevent mistargeting of credit has led to a sharp reduction of agricultural credit operations in the Bank's portfolio and partly explains the decline in agricultural lending by the Bank. Other causes explaining the failure of the integrated rural development approach include (1) adverse policy environment—projects were often pursued in adverse policy environments for agriculture as a whole as well as for the small-scale sector; (2) lack of government commitment—governments did not provide the counterpart funding agreed on during negotiations and required for implementation of the programs; (3) lack of beneficiary participation in technology development—the programs were often designed in a top-down manner and beneficiaries were not given any authority for decision making or program execution; and (4) complexity of coordination—projects were complex in their sectoral coverage and often exceeded local institutional capabilities or, in a futile attempt to avoid the institutional problem, were implemented by "autonomous" management units.

The coordination problem emerged as a consequence of delegating subprogram execution to government bureaucracies or parastatals that were typically highly centralized and had their own objectives. Many of these bureaucracies and parastatals were out of touch with beneficiaries, who could much more easily have coordinated the relatively simple task at the local level (Binswanger 1994). Therefore, the general lesson was that activities should be planned in an integrated manner but should be implemented (or preferably contracted) by line agencies, each working at the pace permitted by its capacity. Another lesson was that projects were designed to supply remedies for supposed—but not proved —unmet effective demands and that the nature of public good was defined too broadly rather than along the lines indicated for research, extension, and rural finance.

As a result of the failure of integrated rural development, the World Bank has retreated from the ambitious rural poverty alleviation agenda of the 1970s and now supports subsector-specific programs and projects. Based on what was learned from the failure of integrated rural development, a recent agricultural strategy paper for the LAC region (World Bank 1993a) formulates an approach that is packaged as nationwide subsectoral operations aimed at achieving the intended institutional development impact. For agriculture, this entails an end to integrated rural development or "area development" projects as such and the application of five strategic themes by the line of agencies responsible for research, extension, and credit enablement. The five strategic themes are (1) continued improvement and fine-tuning of the policy framework; (2) continued improvement in the allocation and quality of public expenditure; (3) development of banking and capital markets; (4) reversal of the degradation of natural resources (soils, forests, and water); and (5) revitalization of technology generation and dissemination.

We should note that poverty alleviation is not included explicitly as one of the five strategic themes on the premise that poverty alleviation will be adequately addressed if the above themes are properly pursued. The Bank expects that a favorable policy framework for agricultural growth will increase income opportunities (regardless of enterprise size), expand the frontier within which private suppliers of agricultural goods and services find it profitable to operate, and increase demand in rural labor and land markets. The Bank also expects poverty reduction to occur from lower food prices, as protection is removed from domestic production of food staples. The Bank asserts that many small farmers, quite apart from the landless, are net consumers of food staples. Therefore, it is anticipated that farmers who are net suppliers of food staples will stand to lose, unless they can increase their productivity or switch to other commodities. According to the Bank, the net suppliers are usually not poor. Nevertheless, it is recognized that poverty reduction in some countries, or in pockets in some areas, will merit more specific attention.

Beyond development of competitive banking and capital markets, the Bank is currently focusing on enabling small farmers to become creditworthy through a proved track record of repaying debt (rather than supplying credit per se).[29] The Small Farmer Services Project in Chile is a good example of the new style program. This project includes land titling, technology transfer, and a directed line of credit with a time limit on any individual's eligibility. Beyond the limit, the individual will have to go elsewhere for credit but will be armed with a land title as well as a debt repayment record acquired during the eligibility period. It is expected that this line of intervention—together with the removal of restrictions on rural financing institutions (as is being examined in projects in Ecuador and Paraguay) and transactions cost subsidies (paid as a lump sum per loan rather than a percentage of the amount of the loan) to encourage new entrants into rural financing and lending to small farmers—will work toward rural poverty alleviation. The Mexico Rural Finance Project is another example of this approach.

The Bank also is engaged in the revitalization of technology generation and dissemination through what it calls pluralistic participation (World Bank 1993a). "Pluralistic participation" refers to the development of national research and extension systems involving several actors rather than relying exclusively on the public sector. To promote pluralistic participation, a recent World Bank study outlines options in the research domain derived from experience in Spain, Chile, Colombia, and elsewhere (McMahon 1994). Recommendations include setting priorities in terms of the public good and establishing a mechanism for contracting out research work through a competitive bidding process in order to tap into, among others, the research capacity of the universities and NGOs. For example, Bolivia's Agricultural Technology Development Project established the Consejo Nacional de Investigación y Extensión Agropecuaria to coordinate more effectively the research and extension activities of various national, departmental, semiprivate, and private agencies and to promote agricultural technology development and transfer through such means as the subcontracting of research to the Centro de Investigación Fitogenética Pairumani (a consortium of private foundations). Mexico's Agricultural Technology Project also provides technical support for a shift toward greater private sector participation in agricultural research and extension, including assistance in the reformulation of the seed law to promote greater private sector research. Colombia already has some experience in pluralistic participation, and Ecuador will soon begin experimenting with this approach through upcoming Bank operations.

In extension, pluralistic participation envisions withdrawing public expenditure from areas where private practitioners are already working (or could work) and contracting service delivery in areas where public financing remains justified. Some positive experiences in this direction come from projects in Chile, Colombia, Costa Rica, Mexico, Argentina, El Salvador, and Nicaragua. In these projects ministerial provision of services has been replaced, and more efficient practices have been introduced—such as voucher systems and contractual private sector or NGO service provision (Wiens 1994). In addition to contracting service delivery externally, projects in Chile and Costa Rica contain pilot programs for moving recipients of free extension into the commercial market for farm management advice. With these projects (and a few others now under preparation in Ecuador, Jamaica, and Bolivia), the Bank expects that improved allocation and quality of public expenditure will improve the human capital of the rural poor. Together with targeted investments in transport, electric power, nutrition, health and sanitation, and education (and by focusing on genuine public good), the Bank expects to shift the content of research and extension toward small farmers' concerns.

The World Bank also chairs the CGIAR system and supports international agricultural research through grants amounting to about forty million dollars per year.[30] Even though these grants are important complements to those made by other donors, the total amounts are too small in relation to the growing demands on the CGIAR system to preserve germplasm resources and to increase yields, address environmental issues, and promote modern biological research for the benefit of developing areas. As a complement to the CGIAR mechanisms that provide funding to international agricultural research institutions, the Bank has recently supported several new research funding initiatives to promote stronger links among the private and nongovernmental sectors, international centers, and national agricultural research institutes in the developing and developed countries. Such funds are being used to finance research proposals submitted on a competitive basis. International centers, private companies, universities, foundations, NGOs, and national research institutions are being encouraged to submit proposals, individually or jointly, for "public good" research—including agricultural research, extension, and technology development (World Bank/AGRTN 1995b).

The World Bank also has supported a variety of biotechnology initiatives in Brazil, Mexico, and Argentina since 1985. Modern biotechnologies range from relatively straightforward and inexpensive procedures of tissue culture to advanced applications of molecular biology, including genetic engineering. Together, these new techniques provide powerful new tools for agricultural research and technology generation. Through Brazil's Third Agricultural Research Project, for example, the Bank is supporting efforts to develop and disseminate environmentally sound agricultural production technologies for the northeast and

Amazonian regions, two of the country's most economically and eco-logically depressed areas. Support for biotechnology is being provided in a cross-cutting research support program at the National Center for Genetic Resources and Biotechnology (CENARGEN). The commod-ity focus is on beans, cassava, oil palm, rare cattle species, and other commodities of economic importance to resource-poor small- and medium-scale farms in the target agroecological regions. The program undertakes upstream molecular and cell biology R&D, maintains gene banks of valuable plant and animal species, and undertakes studies on biological pest control and biofertilizers.

Applications of biotechnology do not always have to be expensive, sophisticated, or capital-intensive. Farmers and local technicians can be trained in skills required to handle biological materials, and these skills can be used in production of biofertilizers, biopesticides, and mi-cropropagated plantlets. In Colombia, for example, farmers are pro-ducing biopesticides on-farm from insect pathogenic fungi grown on rice substrate. The product is formulated and used locally to control plantain, coffee, and potato pests. These prospects should be viewed realistically, however, in light of the constraints on resources, institu-tions, and policies that may be particularly acute in some countries. Many countries have limited scientific infrastructure, human resources, and technology delivery systems, and thus, limited capacity to translate the development of biotechnology products into farm-level benefits. A recent example dealing with strengthening capacity for biotechnology development and dissemination is the World Bank's assistance to Mex-ico. This assistance supports the start-up of the Mexican Institute of Industrial Property and aids in building institutional capacity for the efficient administration of Mexico's 1991 extension of intellectual prop-erty rights legislation, which includes patent rights to biotechnology products (World Bank/AGRTN 1995a).

Beyond revitalizing technology generation and dissemination and in-creasing access to capital, redistributing assets in the agricultural sector (particularly land, but also the knowledge to use it efficiently and sus-tainably) has long been viewed by the Bank as critical for the alleviation of rural poverty. A large array of studies (summarized in Binswanger, Deininger, and Feder 1995) has shown that larger "commercial" farms are frequently less efficient than family farms in terms of value added per unit of area. The World Bank's commitment to land reform has been reaffirmed in a recent Bankwide agricultural strategy paper (World Bank 1993b), which argues that the demise of the Cold War, the fiscal unsustainability of large-scale commercial farm sectors, and new approaches to land reform may finally open up opportunities in this area. The establishment of security of property rights and land tenure and the development of an active land market are objectives pursued by several Bank-sponsored programs, with more being developed (Wiens 1994). Nicaragua's Agricultural Technology and Land Management Project, Paraguay's Land Use Rationalization Project (I–III), and Bolivia's National Land Administration Project, for example, are instances where cadastre, land titling, and land registration are being addressed.

The Bank also is currently promoting "market-assisted land reform," whereby small farmers or farmer groups are provided with a partial grant and a mortgage credit to buy the land of their choosing. Since land prices reflect nonfarming benefits or specific privileges such as tax advantages or credit subsidies commonly acquired by large landowners, the price of land usually exceeds the capitalized value of unsubsidized agricultural profits. For land redistribution to work, this difference will have to be met by a grant. The grant will have to be very large unless the distortions underlying the nonfarming benefits obtained by large landowners are eliminated. Preconditions for successful land redistribution are indeed considerable.[31]

The above initiatives suggest that support for rural poverty reduction has become highly selective within the Bank's program, since it has been impossible to support the full array of interventions required for successful rural poverty reduction, particularly in risk-prone areas. Current Bank-supported area development projects in LAC deal with fewer subsectors than in the past. As a result, agriculture's direct contribution to poverty alleviation is expected to occur through application of the five strategic themes outlined above by the line agencies responsible for research, extension, and credit enablement. By withdrawing from integrated rural development, the Bank has left the unanswered complexities of successfully implementing rural development and poverty reduction programs and has put them in the hands of the country governments. Yet these challenges have not disappeared just because the Bank has withdrawn from them (Binswanger 1994).

After the end of integrated rural development, countries and multilateral lenders (including the Bank) also have proposed other kinds of rural poverty reduction initiatives. These initiatives center on the decentralization of decision making and implementation and that feature small-scale, community-based works with attention to sustainability and an increase in productivity. Decentralized approaches entail serious problems of their own, and there is a compelling need to optimize designs to provide the right incentives and to strengthen local institutions so that effectiveness in responding to community demands can be matched with efficiency in implementation (Wiens 1994).

An interesting evolution of rural development programs has taken place in Mexico and Colombia and, recently, on a pilot basis in Brazil. Programs have evolved into matching-grant mechanisms for rural municipalities or districts or for poor beneficiary groups without necessarily abandoning the earlier multisectoral approach. Within these programs, significant decision-making power over project funds is delegated to municipalities and beneficiary groups through such mechanisms as municipal funds. Within certain budget limits, the municipalities can choose from a menu of poverty-reducing community projects. Project selection takes place according to rules that increase the transparency of decision making to the ultimate beneficiaries and assist in proper targeting to the poorer groups.

Some other countries have recently gone much further with administrative and fiscal decentralization of rural development. For example, Colombia has gradually, and fairly carefully, transferred additional fiscal resources to municipalities, much of it earmarked for health and education. It has also strengthened central government matching-grant systems such as the Integrated Rural Development Program (DRI) and a social fund program. These changes are still being implemented. Considerable administrative deconcentration and delegation of functions has occurred in Mexico under various public sector reform programs. These initiatives have been associated with greater revenue sharing for poorer regions and the development of sophisticated matching-grant mechanisms to lower jurisdictions and community groups. As a result, considerable improvements have occurred in productive and social infrastructure, water supply, and sanitation (Binswanger 1994; Cernea 1992).

Through social or municipal funds programs, the Bank is trying to improve accountability to the poor by encouraging openness and transparency, such as representation of small farmers, women, and rural workers on boards of research stations, supervisory committees of extension systems, or on land or labor committees that deal with such issues in rural areas. In Mexico, for example, all decisions for fund allocations must be taken in open assemblies at the municipal level, and a proportion of the funds must be allocated to outlying communities, which usually are poorer than the municipal headquarters. In Brazil, a special municipal council has been created for the allocation and administration of funds. This ensures adequate statutory representation of poor rural communities in these nonelected bodies (Binswanger 1994).

For the past few years, the Bank has also made loans for projects that directly address resource degradation issues. Within the agricultural sector, these include natural resource monitoring and assessment studies; watershed management and reforestation; land-use planning and management by geographical or socially operational units; improvement of irrigation technology and water management; soil and water conservation; and the development and dissemination of improved farming practices.[32] Recent examples of projects that address natural resource degradation are found in Bolivia and Colombia. Bolivia's Agricultural Technology Development Project is establishing a research program for pasture management and crop production to prevent wind erosion problems in the altiplano. The project also supports research on varietal assistance and biological control of pests and diseases, to obviate the need for the heavy use of toxic agrochemicals. In Colombia, the Rural Development Investment Project is cofinancing watershed management and environmental protection activities in the Andean zone, including protection of soils and vegetation in critical catchment areas, promotion of agroforestry and soil conservation techniques, and support of training and community organization programs for watershed management and environmental protection initiatives.

The World Bank has had some experiences supporting integrated soil and pest management in Brazil and Nicaragua using participatory approaches. The Bank became involved with farmer organizations in 1989 in the first of a total of four Brazilian land management projects in four states in the country. The cost of the first project was 138.3 million dollars; the Bank provided a loan of 63 million dollars. Incentive funds, an important part of launching integrated soil management, have totaled 18 percent of the total cost. The first project developed and implemented a land management plan. Microcatchment plans, generated through the consolidation of individual farm plans and neighborhood plans, were approved by local microcatchment committees made up of area farmers. These plans and funding requests were then submitted to municipal committees on which farmers, extensionists, the private sector, and local authorities were represented. Finally, the requests were submitted to regional and then to state-level committees. A soil conservation fund was created specifically to provide farmers with part of the cost of the new equipment needed to upgrade land management and to reforest areas needing long-term protection. The rationale for such funds was that part of the benefits of these investments were external to the farmers' land. The proportion of costs covered by the fund decreased progressively. The hierarchy ensured, through peer monitoring, that the use of resources from the fund was equitable and that the process was transparent (World Bank/AGRTN 1994a, 1994b).

Other interesting soil and water conservation projects have taken place in Paraguay, Panama, Brazil, Mexico, and Chile (Wiens 1994). Paraguay's Natural Resource Management Project is an example of a systematic, step-by-step approach starting from sector work and progressing to staged implementation of soil conservation. Panama's Land Management Project (I and II) and Brazil's Land Management and Soil Conservation Project also are examples of soil conservation work supported by the Bank. In Mexico, the Irrigation and Drainage Project and three follow-up projects reflect a similar comprehensive and coordinated approach to water management problems. Together with Chile's PROM Project and anticipated activities in Peru, these constitute examples of project experiences where transfers of infrastructure ownership and management to water users are among the issues being addressed.

Forest and biodiversity conservation initiatives have been relatively difficult start-up areas in LAC, given the lack of clear official policy and problems with institutional capacity (Wiens 1994). However, there are several positive experiences in Colombia (Natural Resource Management Project), Mexico (the Lacandona component in Mexico's Decentralization and Regional Development Project I), Honduras (Agricultural Sector Adjustment Project), Bolivia (Natural Resource Management Project), and Argentina (Agricultural Services Project). With Global Environment Facility (GEF) financing, the Bank also is expanding its biodiversity portfolio by promoting the establishment of protected areas in Ecuador, Bolivia, Peru, Brazil, Mexico, Guyana, and Venezuela and is upgrading the management of these areas. There is, however, much to be done in devising participatory ways that local agricultural knowledge can facilitate project conservation and NRM efforts.

Toward a Participatory Strategy in Risk-Prone Areas: A New Paradigm

The major question now facing policy formulators and implementers in relation to poverty reduction and sustainable natural resource management in LAC revolves around where to direct efforts and invest scarce resources—in the high-potential areas? or in the low-potential, risk-prone zones? Significant disagreement on what is required to boost production in the areas of highest potential does not exist, but there is little consensus on the other 80 percent of cropped areas worldwide. Some argue that the best policy to pursue in vulnerable, low-potential areas is to relieve population pressures by encouraging massive out-migration. Although in the long run out-migration may be an answer for some traditional areas, merely transferring poverty and population pressures to urban areas and forest frontiers cannot be an acceptable solution to rural problems. Failure to address problems within the traditional areas themselves will only intensify the problems or shift them to cities, as the rural poor exhaust natural resources and move to join the urban poor.

A large proportion of LAC's poor live in the low-potential, risk-prone regions, and thus, efforts aimed at alleviating poverty would most logically start there. The acute problems found in risk-prone areas, however, pose serious difficulties to policy development and implementation. Such areas have limited potential for growth, are often densely populated for profound historical reasons, have still increasing rural populations and still degrading natural resources, and are inhabited by a relatively large number of unskilled and unhealthy people. In addition to urban bias in the provision of education and health services, there has been a further bias within rural areas against the more marginal lands where traditional agriculture tends to concentrate. Under these conditions, farmers are likely to plunge deeper into poverty, either as smallholders lacking support services and excluded from participatory development processes or by becoming displaced and joining the swelling ranks of urban slum dwellers. Clearly, these biases need to be addressed through specifically targeted rural poverty reduction interventions. Although some poverty alleviation initiatives not related to agriculture per se (such as improvement of the human capital of the poor through investments in education and health) are important elements of a strategy for risk-prone areas, a significantly greater number of initiatives are still required for alleviating poverty, improving natural resource management, and promoting rural development in such areas.

More attention must be paid to areas with fragile ecosystems and large numbers of poor people, and further improvements also must be pursued in the long-term agricultural productivity of high-potential areas in order to meet the needs of rapidly increasing urban populations. Policy makers have to explore the scope for assisting low-potential areas in diversifying production into goods demanded by adjacent high-potential areas and, thereby, benefiting from their growth. Difficult trade-offs do arise, however. Agricultural growth, for example, can usually best be achieved through investments in the highest-potential regions, whereas rural poverty and resource degradation problems are often located in low-potential regions. Hence, public investment in traditional agricultural areas may have high opportunity costs relative to the areas of commercial agriculture where most agricultural growth has occurred. But efficiency trade-offs may be justified in terms of the greater impact on poverty and environmental degradation that can be achieved in traditional areas. Therefore, although in the short term alleviating poverty and improving natural resource management may not necessarily be linked to promoting significant agricultural growth in traditional areas, in the long run poverty alleviation is necessary for achieving sustainability in these areas.

It is evident from the above discussion that the focus of agricultural development must be broadened to include intersectoral linkages influencing employment generation, poverty reduction, and related issues of population growth and expulsion/attraction forces driving rural migration to the forest frontier and urban areas. The investment allocation procedures of donors and governments generally fail to consider the broader questions of how investment in agricultural research and extension, health, education, urban services, and so on may influence decisions on out-migration and resource use in a particular region. Agricultural growth creates powerful multiplier effects on the rural nonfarm economy that enables farm households in degraded lands to mobilize capital and labor for farm investment and rehabilitation, income diversification, evolution of property rights, and infrastructure investments leading to additional employment and earning opportunities for the rural poor (Scherr and Hazell 1993). Off-farm employment, including income-earning activities in rural industry, services, and marketing, is an important component of the survival strategies of the rural poor and hence should be an important focus of antipoverty efforts. Given its crucial importance for overall sustainability, the policy framework for agricultural development must be broadened to incorporate the relationship between geographic and intersectoral allocation of public expenditures influencing employment, poverty, and migration.

Even with the benefits to be gained from economic liberalization, the increasing emphasis on structural adjustment is, in the short term, likely to increase the number of vulnerable and at-risk people in the region. As the state withdraws from organizing rural economic life, a vacuum is left that is not entirely filled by the private sector, as is generally believed. Interactions between public sector agricultural research systems, farmers, private companies that conduct research, private enterprises in food processing and distribution, and NGOs will need strengthening, to assure relevance of research and appropriate distribution of responsibilities. The ability to engage local government, the private sector, and NGOs in addressing problems of technology development and dissemination will largely define the future scenario in the region. This will make government policies, research priorities, and public sector spending more effective and responsive to the resource constraints and local ecological conditions faced by farmers while providing a better foundation for interaction between government and civil society.

To ensure participation in technology development and agricultural research and extension, more resources must be allocated by governments and their donor partners to promote intensive consultations with prospective beneficiaries. Technology development must be not only on-farm and farmer-managed but also participatory, in order to draw on local knowledge and meet farmers' needs, opportunities, constraints, and aspirations (Farrington and Bebbington 1994; Bebbington et al. 1993b; Bebbington 1995; Chambers and Pretty 1994). Enhancing the role of farmers in local analysis, priority setting, experimentation, and other research and extension activities should be important elements of a demand-responsive strategy for dealing with rural poverty and NRM problems in risk-prone areas.

Supervision in the field can yield insights on many issues, such as the desired properties of small irrigation works, desired features of soil and water conservation methods, improved research on diverse cropping systems and long-term viability of farming systems, better integration of livestock and green manures into farming systems, and desired training in activities performed by technicians. Such field knowledge, if reflected in a project's design, can go a long way toward ensuring a convergence of interests around the project. This will function to produce commitment and the sense of ownership by stakeholders without which project success is unlikely.

The relative lack of public and private sector investment in the infrastructure and human resources of many traditional areas in the region has inhibited the full development of the potential of agriculture and natural resources. This lack of attention has left these areas impoverished, isolated, and economically depressed. To reverse this neglect, governments in the region will need to allocate to traditional areas the resources necessary to develop and maintain productive infrastructure and services, including crop storage and processing facilities, irrigation, roads, telecommunications, financial services, and the supply channels for inputs.

In many countries, the potential multiplier benefits of agriculture are constrained by investment codes and related legislation that discriminate against small, rural, nonfarm firms. These policies need correcting. The rural nonfarm economy (as well as rural-urban linkages generally) can be strengthened by removing institutional barriers to the creation and expansion of small-scale credit and savings institutions, and by making them available to small traders, transporters, and processing enterprises. The experience with World Bank–supported projects is that such reforms have been effective in many countries in helping the poor to face risk and to generate more income (Binswanger 1994).

Governments in the region must improve the quality of education, health care, and sanitation in traditional areas. Productive and social sectors are synergistic, not competing. Just as investment in health and education can increase productivity and help achieve economic goals in traditional areas, investment in the development of agriculture and natural resources can help achieve social goals in these areas. Governments will need to revitalize local government in traditional areas and create an enabling environment for local institutions to identify, develop, and maintain new infrastructure and services. To improve efficiency, governments will need to recover costs through user fees, identify and select projects with full involvement of farmer communities based on careful evaluation of potential demand for services, and involve NGOs and private contractors in executing projects.

Incentives and regulatory policies will need strengthening to compensate for externalities related to natural resources. Over time the nature of such policy measures will vary across countries, but they are likely to include policy reforms dealing with water allocation mechanisms and watershed management; exploitation of lands and forests resulting from free access; determining and enforcing property rights including land, water, and forests; and correcting distortions in input and output markets, asset ownership, and other institutional and market distortions that are adverse to the poor. Regulations will be necessary where incentives are unlikely to achieve social objectives. However, policy makers must be aware that regulations contradicting the survival strategies of poor farmers are unlikely to be successful, simply because they will be difficult or impossible to enforce in traditional areas. Finally, serious commitment to promoting equitable access to land, water, and capital in traditional areas will be crucial for achieving sustainable development and poverty reduction in the region.[33]

An important aspect in the development of traditional areas is the strengthening of political organizations such as indigenous, farmers', and women's groups. These have endured historical patterns of exclusion that persist to this day. The violence endemic to the region compounds these difficulties. Although in many cases this violence is born of long-standing social, economic, and political inequalities, in others (such as the Andean region) it reflects the rising influence of the illegal drug trade. This influence has infiltrated the societies of almost every country in the region, distorting the economic and social development of rural areas and undermining citizens' confidence in public officials and, indeed, in the entire justice and political system (Garrett 1995).

After decades of advocacy and support of the smallholder sector by scholars, bilateral donors, NGOs, private voluntary organizations, and human rights organizations, there is wide recognition of the need for farmer empowerment. As long as small farmers in risk-prone environments continue at the periphery of the political process, and governments continue to favor the politically powerful urban population, these poverty-stricken and ecologically vulnerable rural areas will remain ignored—except as pools of labor or as target zones for welfare transfers. Consistent with human rights and democratic governance, it should be made legally and institutionally easier for resource-poor farmers in traditional areas to group together in order to gain their own political voice and clout.

The weak legitimacy of many governments in the region hinders policy makers from formulating and implementing rural strategies capable of mobilizing the population. Hence, mobilizing effective support for Latin American countries that are moving in the right direction and that can serve as models to other countries in the region is just as urgent as reducing support for those that have favored large farms and corporations to the neglect of small farmers and the rural poor. Although officially recognized as an essential means of fostering economic development and socioeconomic stability, improvement in income distribution is still not operationally relevant to policy decisions in most countries in the region. The difficult choices and decisions based on a long-term view that addresses the costs and trade-offs of resolving these conflicts still have not been made. Nevertheless, we are convinced that, at least for now, the World Bank is one of the few institutions with the influence and resources to stimulate action in the region and to carry through a consistent strategy for the risk-prone environments.

PART II

Theoretical and
Practical Issues

2

Rural Development and Indigenous Resources

TOWARD A GEOGRAPHIC-BASED ASSESSMENT FRAMEWORK

Bruce A. Wilcox

Geographic or map-based data and their analysis have proved useful, if not essential, in many areas of science, including research on natural resources for their conservation and management. Such approaches, incorporating indigenous characteristics of geographically delineated regions, are proving valuable for assessing conservation and development priorities. The World Bank's Ecoregion Project for Latin America and the Caribbean (Dinerstein et al. 1995), which recently provided the basis for regionwide participatory investment priority setting (BSP 1995), is a good example. Based on geographic information systems (GIS) technology, the visually appealing and familiar appearance of maps makes otherwise complex and unwieldy data accessible to a broad group of users and stakeholders. With the increasing availability of GIS technology, this form of information is providing the basis for an effective means of cross-cultural communication.[1]

This chapter discusses the application of geographic data to the problem of rural smallholder development, framed in a systems perspective

Bruce A. Wilcox

in which the biophysical, sociocultural, and economic attributes of a rural area are seen as interdependent and influenced by those of the urban-industrial sector. The chapter describes how this systems analytical view of sustainable resources management and development— together with the data on indigenous resource values and factors indicative of human pressures causing environmental degradation—is potentially applicable to an assessment of rural development needs and opportunities. Indicator methodologies employing these data in a GIS format are described as the principal means of assessment.

Of two indicator approaches to be discussed, the first is based on the valuation of ecoregions in the Latin America and Caribbean (LAC) region, used in the above-mentioned priority setting exercise. The results suggest that, although differing perspectives and the pressures created by an increasingly dominant urban-industrial sector mitigate against smallholder development, valuation of indigenous resources provides a basis for reformulating current policies. The second approach, based on an environmental policy indicator model, employs additional kinds of georeferenced data as a means of developing pressure, state, and response indicators analogous to the environmental policy indicator approach adopted by many OECD countries (see, for example, Adriaanse 1993). A variation of this approach, applied to local socioeconomic or ecologically defined regions, uses data on population and land-use change—along with measures of biotic integrity and ecosystem health —to assess the threat of environmental degradation (ISD/RIVM/ WRI 1995). The method is applicable to identifying the risk-prone rural environments discussed by Pichón and Uquillas (this volume). Combined, the two approaches illustrate the value of geographic data and their appropriate analysis for providing a rational basis for policy development and decisions that foster smallholder development within a sustainable resources management context.

Framing the Problem

The focus of this volume, the use of the natural resources of a given area to sustain livelihoods (Pichón and Uquillas 1995), can be framed using a holistic systems-analytical approach for defining and conceptualizing sustainable development. This approach recognizes that any system consisting of human-built or natural elements or a combination of the two can be viewed as hierarchically structured and consisting of interacting components (subsystems). In this view, for example, a household economy exists within a village economy, which exists within the economy of a local region, which exists within a national economy, and so on. A similar structure can be described for biophysical systems or biological diversity generally, as well as sociocultural systems.

For the world as a whole or any given region, country, district, or village, the subsystems can be visualized as interdependent (or integrated) and in combination constitute a single, whole system, which can be called a socioecological system (Gallopín, Gutman, and Maletta 1988; Shaw et al. 1992). This inclusive, integrated system is arguably that which should be addressed when confronting the problem of sustainable development. The interaction of these subsystems and their relationship to development is shown in figure 2.1.

It follows that a socioecological system, described on a scale appropriate for rural development assessment purposes (such as a region or district within a country), both contains and is contained by other systems at successively smaller or larger scales (respectively) with which it interacts. This has critically important implications to natural resource management in general, and rural smallholder development in particular (which is, by definition, an integral part of the former).[2] Almost any conceivable rural socioecological system is part of a larger socioecological system and subject to its ecological, sociocultural, and economic influences, and vice versa. Tensions resulting from the negative influences of one system on another can help define the challenge of rural smallholder development, as well as sustainable development more generally.

Such influences or tensions existing between adjacent rural systems or between one local system and that defined by an entire district, for example, are an important consideration in development planning. However, certainly the most significant negative influences on rural systems that generally work against smallholder development stem from the interaction between rural systems and the increasingly dom-

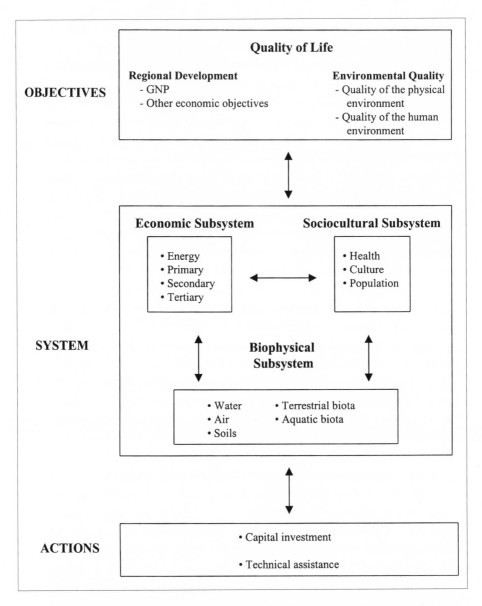

Figure 2.1. A Socioecological System and Its Development (based on Gallopin et al. 1989)

inant "urban-industrial system," which can be taken as a whole to include the global urban-industrial system or the urban-industrial systems defined regionally or within a country. Smallholders and their rural environment, though geographically distinct, are significantly influenced by the urban-industrial sector, considered on any scale.

Some fundamental distinctions exist between the two production modes and sectors. In contrast to landscapes dominated by urban-industrial modes of land use and production, in landscapes where natural resource–based land-use systems are the dominant production mode, traditional cultural and indigenous biological features often remain highly specific to place or geographic area. Terms such as "bioregion," "ecoregion," or simply "region" are used by natural and social scientists to define areas according to their natural and cultural (or socioeconomic) features. Particular zones represent different attributes in soil, topography, climate, and biota as well as in sociocultural aspects, all of which have been determined by a combination of historical factors and contemporary environmental conditions. These natural and cultural attributes together with external influences (often constraints) determine the region's development needs and capacity—including the opportunity to improve natural resource management by addressing, for example, problems of poverty and environmental degradation through smallholder development.

Two sets of dichotomies can be seen when comparing urban-industrial and rural modes of production and development. These help explain some of the major tensions and problems arising from their interaction. First, from a biophysical subsystem standpoint, the urban-industrial systems are driven primarily by fossil fuel energy and controlled by human-designed processes, whereas the rural systems are primarily driven by solar energy and controlled by natural or human-modified ecological and evolutionary processes. Second, from a sociocultural subsystem standpoint, societal values and perceptions in the urban-industrial sector derive from the perspective of the dominant production mode (urban-industrial), whereas those of the rural sector are influenced primarily by a view of economic production founded on the indigenous resource base of their local rural environment. Gadgil (1993) describes a corresponding sociocultural distinction, referring to people or societies whose economic needs are met in an ecologically diffuse sense by the larger system as "biosphere people," and to those who associate their economic welfare with the rural natural resource–based system as "ecosystem people." As the following suggests, these two dichotomies must be dealt with for rural smallholder development and, more generally, sustainable development to be achieved.

Considering those influences on the biophysical subsystem (figure 2.1) from the standpoint of the whole system, biosphere and ecosystem people both often produce and suffer from the effects of each other's modes of production. For example, some rural production modes (such as those involving forest clearing) may negatively impact the urban-industrial environment through their effects on watershed capacity. Just as often (or more so), the negative impact is in the other direction. Industrial modes of production often result in excessive pollution loads, negatively affecting rural environments and production capacity. Acid precipitation, ozone production, and various kinds of water pollution from industrial activity are severely impacting the biotic integrity and ecosystem health of rural areas in many developing countries, thereby degrading the environment and undermining the capacity for economic development.

The urban-industrial sector's emissions of carbon dioxide, perhaps the most troublesome form of pollution globally, reveals a more complicated situation. The excessive production of carbon dioxide by industrial production modes—or that carbon dioxide production caused directly or indirectly by deforestation because of the economic influences of the industrial sector on the rural sector—has resulted in policies, promulgated by biosphere people, that affect the rural production modes of ecosystem people. That is, the consequences of rural carbon release or sequestration have been a significant concern among biosphere people who, in fact, are the ones who bear nearly all of the responsibility for the historic increase in atmospheric carbon dioxide. Indeed, global carbon balance is hardly a priority for the average campesino or traditional cultivator, most of whom have more immediate economic concerns. Still others, particularly in Latin America, have been pushed onto lands away from coastal lowlands likely to be inundated by a rising sea level. Thus, while in many instances the rural sector (or the ecosystem people) might be considered downstream from the immediate negative effects of urban-industrial production modes (of the biosphere people), the consequences, viewed on a larger scale, may weigh more heavily on the globally dominant sector. In fact, biosphere people may ultimately suffer more from rural environmental degradation—for this and other reasons that will be elaborated shortly.

For a sustainable global socioecological system to be realized, the urban-industrial and rural smallholder sectors obviously must coexist on the basis of mutually beneficial and ecologically sustainable economic activities. The current situation is that the actions and pressures of economies driven by biosphere people, including the recruitment of ecosystem people to become biosphere people, generally underlie deforestation and rural environmental degradation in Latin America. This pattern has been extensively documented for Latin America by numerous studies (see Painter and Durham 1995 and Norgaard 1994 for broader theoretical explanations).

Although the conflicting sociocultural and biophysical consequences are mutually disadvantageous, those of an economic nature are primarily unidirectional in their negative effect on smallholder development. Reconciling conflicting influences of the economic subsystems of the urban-industrial and rural smallholder production modes poses perhaps the greatest challenge to rural smallholder development and, possibly, to global sustainability. Production of food or other commodities by industrial production modes in the agriculture, forestry, or even the fishery (commercial aquaculture and mariculture) sectors necessarily represents a major impediment to rural development. The industrial sector's capacity to externalize ecological and social costs (often through their "export"), economies of scale, and market access contributes to a decided competitive disadvantage for rural smallholders. The emergence of the global exchange economy has tended only to worsen this situation. Thus, given the current set of conditions, the economic subsystem driven by biosphere people has overwhelming downstream effects on that of ecosystem people. This can reasonably be remedied only through policies that minimize the capacity of industrial production to subsidize its "efficiency" with externalities, combined with approaches to rural development that seek a comparative advantage based on the most significant characteristic distinguishing rural socioecological systems, that is, their indigenous biophysical and sociocultural resource base. This provides the most hopeful solution to smallholder development, as both beneficiary and benefactor.

Framing Solutions

A cogent and compelling case must be made for reducing "subsidization" of industrial production modes, including those directly or indirectly resulting in rural poverty and environmental degradation. This case cannot be based solely on moral and ethical arguments, which, however valid, have proved time and again inadequate by themselves, given present-day economic and political realities. The concept of indigenous resources provides a basis for building such a case.

For the purposes of this discussion, "indigenous resources" are defined as the sociocultural and biophysical attributes of a site, locality, or region such as an ecoregion. These attributes have coevolved, or at least are preadapted in situ, and represent a form of inherent "capital" or "productive assets," of which the production capacity can be enhanced through internal reorganization, external inputs, or both. It is a given that external inputs of capital or technology cannot be readily substituted for indigenous resources, since by definition they represent complex structures, finely tuned to each other and the physical environment as a result of generations of cultural and biological evolution.[3]

The above case can be made using indicators of direct or indirect economic value, based on indigenous resources attributable to ecoregions in Latin America. Based on the tabulation of values and benefits, many indicators can be seen as coinciding—despite the conflicting biosphere and ecosystem perspectives. The utility benefits that derive from the ecosystem processes and biodiversity characteristic of an ecoregion are of both global and local significance. Those shown in table 2.1 are primarily what ecological economists refer to as indirect functional benefits of ecosystems or ecosystem services. When these utility values are measured or ranked, the same ecoregions tend to exhibit the highest global and local values.

This can be seen in table 2.2 by comparing net primary productivity

Table 2.1. Summary of Biodiversity Utility Value and Benefits

Value Category (measures/indicators)	Benefit (productive or functional)
Carbon sequestration Biomass carbon content Soil carbon accumulation	•Amelioration of global climate change
Genetic resources Centers of plant diversity Origins of important crop species Forest tree genetic resources	•Source of wild species of actual or potential utility •Source of genetic material for improvement of domestic species •Source of domesticated or farmed species
Hydrological resources Major river systems	•Provision of water resources •Regulation and moderation of water flow •Production/transport/deposition of nutrients •Provision of water transport
Productive and protective resources Allochtonous productivity Protection Net primary productivity	•Source of nutrients/nursery for fishery stocks •Protection from coastal/riparian erosion •Assimilation of pollutants •Biotic production potential
Indigenous Resources Areas of indigenous populations	•Source of all economic, aesthetic, and cultural resources for native peoples •Source of traditional knowledge and enhancement of biodiversity
Ecotourism, recreation, research resources	•Source of amenity and tourism value •Source of scientific knowledge
Significant watersheds	•Watersheds providing critical drinking water, irrigation, and hydroelectric needs
Significant productive areas	•Ecosystems whose productive output and assimilative capacity is critical to local urban, peri-urban or a nonindigenous rural subsistence economy
Spiritual or aesthetic values	•Sources of intrinsic value to a local community or communities in a country

(an indicator of the potential productive capacity of the land) with the other categories of utility value. This shows that although biosphere and ecosystem people do not share the same values, more often than not both groups benefit from—and their economic well-being ultimately depends upon—the same indigenous resources. This argument, based on geographically coincident biophysical resource dependencies, can be extended to sociocultural resources. Comparison of biological utility indicators with the number of ethnolinguistically distinct indigenous populations in ecoregions shows that cultural diversity (by this measure) and biological diversity are correlated in Latin America (Wilcox and Duin 1995). These local (or, in the case of large areas such as the Amazon Basin, regional) ecosystem-based societies are in effect the stewards of some of the most significant resources available to global society. Even when local societies are neither indigenous nor traditional and lack knowledge in natural resource management, they are de facto stewards upon whose latent or potential resource management capacity society at large must depend. A most significant category of indigenous resources that are crucial to the urban-industrial sector is that of genetic resources.

It is well known that genetic resources (comprising land races, primitive cultivars, and wild relatives upon which the world's agricultural production is based) ultimately derive from agroecosystems and their associated natural ecosystems, which are inhabited and managed by traditional or indigenous people. It is ironic that the expansion of high-yield strains and high-input agriculture made possible by these resources often contributes to their demise (NRC 1993a). Despite the exceptional efforts of numerous institutions and individuals to collect, describe, catalogue, and store germplasm or otherwise conserve genetic resources ex situ, the amount of genetic diversity so conserved is small by comparison to that which remains in situ. (Ex-situ conservation refers to the maintenance of genetic diversity—typically in an artificial environment such as tissue culture, seed, frozen ova or semen, a zoo, or a botanical garden—rather than the local natural environment where a gene pool or population evolved.) Numerous cases are recounted in which a crop has been saved by an obscure variety, found in situ, carrying a disease-resistant trait. Even though the genetic diversity for modern crop improvement typically comes from an ex-situ repository, the repository represents just the end of a "genetic diversity pipeline" that begins with wild varieties in relatively undisturbed ecosystems and includes the intermediate steps of semidomestication by local farmers. The remarkable capacity of modern biotechnology to manipulate genetic material does not negate the necessity for genetic raw material but increases its value.

The conservation of genetic diversity, as one element of biological diversity in rural landscapes, is—or ought to be—as important a goal of industrial society at large as it is to local cultivators in genetic resource-rich areas. That is, the conflict between modern industrial agriculture systems and traditional local systems should be resolved into an opportunity. Various approaches are being studied and a few implemented, in this regard, involving policies and programs to encourage land-use systems consistent with in situ conservation of genetic resources. Yet, there are no comprehensive regionwide efforts or geographic assessments of the needs and opportunities. The centers of origin and diversity of the world's most important crop plants have been crudely mapped to ecoregions (Wilcox and Duin 1994) to generate indicator values shown in table 2.2. Nonetheless, actual locations and distributions of recognized and important close relatives of these crop plants are generally poorly known and, apparently, only partially sampled for ex-situ preservation.

The limited focus of development policy and activities is remarkable in view of the economic value of these crop plants to society at large. A compilation of the world's most economically valuable crop plants with genetic origins in Central America and the Andes have crop values to the United States alone in the range of between five and ten billion dollars annually (table 2.3). The waning flow of Latin America's genetic resources into the U.S. agricultural production system may not be felt for some time, because of the germplasm reserves stored ex situ. The fact remains, however, that several percent of U.S. agricultural production each year is attributable directly to the genetic diversity pipeline. One calculation for *Theobroma cacao* (the source of chocolate) shows that yield improvements based on wild genetic resources contribute fifty million dollars to annual U.S. production (Prescott-Allen and Prescott-

Table 2.2. Indicators of Biodiversity Utility Value for Ecoregions of the Central America and Andean Region (from Wilcox and Duin 1994)

Ecoregion	Biomass Carbon Content[1]	Forest Tree Genetic Resources[2]	Centers of Plant Diversity[3]	Origins of Important Crop Species[3]	Domesticated Animal Origins[3]	Net Primary Productivity[4]	Indigenous Population
Tropical Broadleaf Forests							
Colombian/Venezuelan Moist Forests	20	25	3	1	1	2200	17
Western Amazon Complex	20	27	5	2	5	2200	86
Northern Amazon Complex/ Guyana Complex	20	33	3	3	4	2200	6
Southern Amazon Complex	14	4	0	2	1	1600	8
Northern Andean Montane Forests	14	4	7	7	3	1600	2
Central Andean Montane Forests	14	5	3	6	7	1600	23
Central American Moist Forests	14	28	4	6	3	1600	27
Mexican Moist Forests	14	2	1	3	2	1600	11
Northern South American Dry Forest Complex	7	8	0	1	0	800	6
Western Andes Dry Forest Complex	7	2	2	1	3	800	3
Cerrado Dry Forest Complex	7	7	1	0	2	800	7
Central American Dry Forests	7	15	0	1	3	800	2
Mexican Dry Forests	7	3	0	1	2	800	2
Conifer or Temperate Broadleaf Forests							
Mexican/Meso-American Conifer/ Mixed Conifer Forests	7	12	2	0	0	1300	23
Grasslands/Savannahs/Wetlands/Shrublands							
Southern Cone Complex	4	6	1	0	4	700	4
Amazonian Savannahs	3	0	1	0	0	900	15
Venezuelan/Guyanan Complex	3	1	0	0	1	900	7
Grasslands/Savannahs of Meso-America	3	0	0	0	1	900	3

Table 2.2. continued

Ecoregion	Biomass Carbon Content[1]	Forest Tree Genetic Resources[2]	Centers of Plant Diversity[3]	Origins of Important Crop Species[3]	Domesticated Animal Origins[3]	Net Primary Productivity[*]	Indigenous Populations
Southern Cone/Cerrado Wetlands	3	1	0	0	1	2000	2
Paramo	1	0	5	1	0	600	1
Puna	1	6	1	0	6	600	14
Xeric Formations							
Venezuelan/Colombian Xeric Complex	1	2	1	1	1	250	1
Pacific Coast SA Xeric Complex	1	3	1	1	3	250	3
Mexican/Central American Xerics	1	5	0	0	1	250	0
Freshwater and Coastal							
Pacific Equatorial Complex	7	0	0	0	1	2000	2
Meso-American Pacific Coast Complex	7	0	0	0	1	2000	2
Gulf of Mexico/Yucatan Basin Complex	7	0	0	0	1	2000	2
Meso-American Caribbean Coast Complex	7	1	1	0	1	2000	3

Notes: Countries include: Central America—Belize, Guatemala, Honduras, El Salvador, Nicaragua, Costa Rica, Panama; Andean Region—Colombia, Ecuador, Peru and Bolivia

[1] Units are in kg/m². Data source Olson, Watts, and Allison, 1983.

[2] FAO 1994.

[3] Updated from original data in WCMC 1992.

[*] Units are in g/m²/yr. Data source Lieth and Whittaker, 1975.

Bruce A. Wilcox

Table 2.3. Important U.S. Crops with Genetic Origins in Central America and the Andean Region

Family	Species	Crop	Economic Crop Value to U.S.A. Import value in Parentheses Avg. annual value 1976–1980 (U.S. millions)
Caricaceae	*Carica papaya*	Papaya	10.3 (1.9)
Chenopodiaceae	*Chenopodium quinoa*	Quinoa	?
Cucurbitaceae	*Cucurbita maxima* *Cucurbita moschata*	Squashes/ Pumpkins	11.8 (11.8) (For Cucurbita)
Juglandaceae	*Carya illinoensis*	Pecan	101.5 (1.8)
Lauraceae	*Persea Ameriana* *Lupinus mutabilis*	Avocado Lupin	93.8 (0.4) ?
Leguminosae	*Phaseolus lunatus* *Phaseolus vulgaris*	Lima Bean Comm. Bean	50.6 (0.04) 516.6 (10.7)
Malvaceae	*Gossypium barbadense* *Gossypium hirsutum*	Cotton	4233 (439.0)
Myrtaceae	*Pimenta dioica*	Pimento	?
Orchidaceae	*Vanilla planifolia*	Vanilla	21.1 (21.1)
Rubiaceae	*Cinchona calisaya* *Cinchona officinalis* *Cinchona pubescens*	Cinchona/ Quinine	37.5 (37.5)
Solanaceae	*Capsicum annuum* *Capsicum frutescens* *Lycopersicon esculentum* *Solanum tuberosum*	Peppers/ Chilis Tomato Potato	158.1 (53.7) (For Capsicum) 1051 (158.0) 1206 (10.2)
Sterculiaceae	*Theobroma cacao*	Cocoa	1016 (1016.0)

Source: Table 7.3, Prescott-Allen and Prescott-Allen 1986.

Allen 1986). The figure for the benefits of wild as well as semidomesticated and domesticated resources for all indigenous Latin American genetic resources to all industrial agricultural economies is certainly billions of dollars annually.

Why in situ genetic resources conservation is not a higher priority of multilateral and bilateral development assistance agencies is unclear, especially when it can be integrated with smallholder development and other environmental conservation needs at almost no extra cost. Possibly, part of the answer lies with the failure of past development assessment approaches to recognize these needs and opportunities. Another factor might be related to the perception of values discussed below.

More comprehensive mapping and listing of these plants would be a first step toward such a priority-setting assessment for multilateral and bilateral rural development assistance. As a secondary factor, higher resolution georeferenced data on indigenous biological and cultural resources for Latin American ecoregions should be compiled and should link detailed sociocultural and economic data with those for specific genetic resources. This would provide an important policy- and decision-making tool applicable to assessing risk to a critically important component of globally significant biological diversity as well as providing opportunities for mitigation through smallholder development.

Similar arguments can be made for the other categories of utility benefits summarized in table 2.1. Each can be framed on the basis of a dichotomy between two contrasting modes of production and sets of system characteristics, including perceptions affecting the values attributed to indigenous resources and encompassing ecosystem goods and services. The possible causes of the dichotomy and the conflicts that result obviously merit much greater consideration in development assistance policy.

The production modes of the urban-industrial sector are based primarily on both natural and human capital that are essentially nonindigenous and nonrenewable. The economic, sociocultural, and biophysical characteristics of urban-industrial systems derive largely from these qualities. On the other hand, rural production modes depend more on indigenous human and natural capital from which their accompanying qualities are derived.[4] These include knowledge and values based largely on a sense of dependence on the land, soil, natural communities, and natural cycles, recognized by some of the urban-industrial sector as biodiversity and ecosystem functions.

The above discussion attempts to frame and operationalize these concepts in reference to land units, such as an ecoregion, whose definition and characterization are made increasingly precise and capable of greatly facilitating cross-cultural communication using georeferenced data. The management, analysis, and presentation of georeferenced data are made possible by recent advances in remote sensing and GIS technology. The framework and the GIS data approach described here represent not only a priority assessment tool but a means of narrowing the "ecosystems value gap" that exists in the world's increasingly dominant urban-industrial society. Unlike indigenous or traditional societies, living in the local ecosystems from which they derive ecosystem goods and services, urban and industrial societies cognitively disassociate themselves from land and its natural communities and cycles. The result is a gap between the actual and the perceived values of ecosystem goods and services (Wilcox 1994).

Smallholder Agriculture: Development Options and Ecosystem Management

The problem of smallholder development—as defined here to include natural resource management—can be viewed from two operational perspectives. One perspective focuses on assisting in the enhancement of smallholder income through higher production; the other on ensuring that the first will result in the net improvement of the overall sustainability of the land resource base. Downstream effects of the global industrial sector aside, enhancement of smallholder agricultural production through conventional agricultural intensification (based primarily on increased external inputs such as fuel, pesticides, and fertilizers) may not be an option for most smallholders, for both economic and ecological reasons. Economic reasons include the scale of operations and access to markets and capital. Ecological reasons include the lack of suitable land for conventional intensification approaches. Smallholders often have access only to lands marginally suitable for agriculture because of inadequate water, soil, or level ground. The frequent occurrence of nutrient-scarce soils in the humid tropics is well appreciated by agricultural researchers.[5] Any comparative advantage sought for smallholder agriculture development typically will be constrained by these factors.

Smallholder development incorporates the broader issue of resource management—including protection of the utility and nonutility values attributed to forests and other natural ecosystems. Although this may, at first, appear an additional constraint, the solutions to smallholder development should simultaneously conform to broader resource management needs. Again, a systems perspective is helpful in framing possible solutions. From such a perspective, the integrated management of a larger land unit—consisting of a mosaic of land-use systems—would be considered from the standpoint of different properties or attributes appropriately adjusted to biophysical, sociocultural, and economic conditions. This requires intensification of information use, rather than intensification of external inputs per se, and more extensive and coordinated management on a regional ecosystemic scale. Although seemingly complex, addressing the problem from this perspective is a requisite part of the solution to smallholder development and, in fact, can be achieved simply by combining approaches to indigenous resource–based smallholder development and approaches for cooperative regional, district, or watershed-scale planning and management.

As an example, one might imagine this dual-level approach to rural smallholder development assessment and planning as applied to a local region or watershed in the humid tropics. Within such an ecosystem, where the problems of rural poverty and environmental degradation are often most severe, a range of land-use systems are possible—each with its own advantages and disadvantages (see table 2.4). The actual suitability of each land-use system depends on highly site-specific or socioecological system–specific conditions. The collective amount (in terms of area covered) of land-use systems, their distribution, and their geographic or spatial association within an ecosystem will influence important regional biophysical characteristics. These include local meteorological and hydrological regimes, biological diversity, and ecological processes (all of which are functionally interrelated). Thus, for any regional ecosystem, the various mixes, patterns, and dynamics of land-use systems will determine its overall functional integrity and, therefore, its capacity to sustain smallholder development as well as to provide locally and globally significant utility values. Various methodologies exist—based on ecosystems science, landscape ecology, and conserva-

Table 2.4. Comparison of the Biophysical, Social, and Economic Attributes of Land Use Systems in the Humid Tropics

Land-Use Systems	Biophysical Attributes					Social Attributes			Economic Attributes		
	Nutrient Cycling Capacity	Soil and Water Conservation Capacity	Stability Toward Pests and Diseases	Biodiversity Level	Carbon Storage	Health and Nutritional Benefits	Cultural and Communal Viability	Political Acceptability	Required External Inputs	Employment per Land unit	Income
Intensive cropping											
High-resource areas	M^{MH}	M	L^{M}	L	L	M^{H}	H	H	H	H	H
Low-resource areas	L^{M}	L^{M}	L^{M}	L^{M}	L	M^{H}	M^{H}	M^{H}	M	M^{H}	L^{M}
Low-intensity shifting cultivation	L	L^{M}	L	L^{M}	L^{M}	H	M^{H}	M^{H}	M	H	M
Agropastoral systems	M	M^{H}	M	M	M	M^{H}	M^{H}	M^{H}	M	M^{H}	M
Cattle ranching	L	L^{M}	M	L^{M}	L^{M}	L^{M}	L^{M}	H	M	L	M
Agroforestry	M	M	M	L^{M}	M	M	M^{H}	M^{H}	M	L^{M}	L^{M}
Mixed tree systems	H	M^{H}	M	M	M^{H}	L^{M}	M^{H}	M^{H}	L^{M}	L^{M}	L^{M}
Perennial tree crop plantations	H	M^{H}	L	L	M	L	M	H	H	M	H
Plantation forestry	H	M^{H}	L^{M}	L	M^{H}	L	M	H	M	M	M
Regenerating and secondary forests	H	H	M	M	M^{H}	L	M	M	L	L	L
Natural forest management	H	H	M	M	M^{H}	L	M^{H}	M^{H}	M	M	M
Modified forests	H	H	M	M	H	L^{M}	M	M	L^{M}	M	L
Forest reserves	H	H	H	H	H	L	L	M	L	L	L

Source: Adapted from NRC 1993b, table 3.1.

Note: The letters L (low), M (moderate), and H (high) refer to the level at which a given land use would reflect a given nature using the best widely available technologies for each land-use system. Superscript letters refer to the results of applying best technologies now under limited-location research or documentation. The systems could have these improved characteristics given short-term (5–10 years) research and extension.

tion biology—that are applicable to the planning and management of regional ecosystems to optimize these needs and values.

Intensification of production by conventional means generally will be challenged, however, by economic or ecological conditions that render these options unsuitable. Those conventional means having the most promise in nutrient-limited, or otherwise marginal, forested or formerly forested areas are those conforming to the basic principles of tropical ecosystems science: maintenance of structural diversity, maintenance of soil organic matter, minimization of soil disturbance, and control of the size and shape of the disturbed area (see Jordan 1985: 148–51). Mixed cropping systems that employ annual and perennial crops, including trees (agroforestry), conform most closely to these conditions. Similar options exist outside humid tropical zones where conditions also may be limiting.

Mixed cropping systems hold the most promise for local societies to exploit indigenous resources with external assistance. Recent successes in dramatically increasing crop yields in Bolivia—based on indigenous knowledge combined with indigenous genetic resources—provide a remarkable example. With the assistance of World Neighbors, Quechua and Aymara farmers established experimental plots using different local and introduced varieties of potatoes and other indigenous crop plants. On the basis of their own statistical analyses, they could adjust cropping practices and local varieties according to their existing knowledge of the varieties, microclimates, and soil conditions in their fields. The resulting yields significantly exceeded those based on cropping practices and varieties being encouraged by agricultural research centers (Ruddell 1995).

Utilizing traditional knowledge of local biological cycles and interactions as well as actual or potential genetic resources as a basis for enhancement of indigenous commodity production represents a potentially significant comparative advantage for smallholders. Perhaps the most economically underexploited resources in this regard are nonwood forest products (NWFP). There is, especially in forest ecosystems, a diverse array of unexploited species and products, many of which cannot be cultivated elsewhere because of mutual interdependencies with pollinators or seed dispersers. Such a combination of conditions provides a comparative advantage for smallholders, assuming they can find a market. The international trade in nonperishable commodities, and particularly in NWFPs, presents attractive possibilities.

Innovation in production methods and crop choice based on traditional and modern knowledge remains a basic challenge. Anderson (1989) presents an interesting model for such innovation based on land-use systems in the Brazilian Amazon. He notes that present extractive forest and extensive agroforestry systems are not sufficiently productive to compete with other land-use systems, such as shifting agriculture and cattle raising. He concludes, however, that by integrating existing systems of extractive forestry, extensive agroforestry, and intensive agroforestry, new production systems that intensify production per unit area could be developed, with minimal reduction in ecological sustainability.

In another encouraging study, based on existing cultivars and the development of new crops, Clay and Clement (1993) describe how "income-generating forests" offer opportunities for smallholders. This report and other recent technical publications describe numerous species with income-generating potential, including specifics on their chemical composition, food value, breeding, planting, harvesting, yields, processing, distribution, and marketing (FAO 1986a, 1986b, 1988b, 1991b). The development of new crops and mixed cropping systems that produce respectable yields requires a considerable investment of time, if not capital. Once these hurdles are overcome, an additional investment is required to successfully market a product. On the positive side, the international trade in a vast array of NWFPs has firm historical roots and might be further strengthened by the growing global exchange economy. In fact, the domination of the international forest products market by industrial wood is a relatively recent phenomenon (Iqbal 1993). The recent explosion of technical information on potential products, their commercial development, and market potential has been accompanied by publications that explore the business management and financing mechanisms for small-scale forest enterprises (see, for example, FAO 1987, 1989).

Smallholder development must be planned and implemented on a regional ecosystem level, taking into account the microclimatic, hydrological, and biological processes upon which these and other kinds of agroecosystems depend for their productivity. Each region has unique indigenous biophysical attributes that can generally be characterized by major ecosystem type as in the ecoregion project. Different regions will have different priorities and require different approaches to smallholder development, natural resource management, and biodiversity protection.

Policy Indicators for Rural Development

The socioecological systems perspective for assessing smallholder challenges and opportunities makes apparent that similar multidisciplinary approaches are needed as both diagnostic and prognostication tools to indicate which rural systems are under the greatest threat and the ameliorative action required. Pichón and Uquillas (this volume) allude to the need to identify "risk-prone environments" that might, in the context of this discussion, be called risk-prone socioecological systems. This sentiment has been echoed by two significant technical publications: the U.S. National Academy of Sciences report on sustainable agriculture in the humid tropics (NRC 1993b), and the more recent International Workshop on Agroecosystem Health with special reference to CGIAR (Nielsen 1994). Both effectively emphasize the need for more and better information on land-use patterns and classification, the need for local knowledge (including information on indigenous production systems), and the need for a systems-based approach.

The reports, along with numerous other recent studies, indicate that digitally stored and managed georeferenced or map information is the essential requirement for assessment and monitoring. Such information, because of the recent dramatic increases in accessibility to computer technology, is empowering local (and even indigenous) people with the information required both to manage their indigenous resources and to defend their land and resource rights (*Cultural Survival* 1995). In addition to empowering the public at large, GIS technology applied to the appropriate data provides information on the status of resources otherwise unavailable to policy makers and planners at district, national, and higher levels.

Among the problems that remain in fully realizing the potential of this information is the problem of articulating operational frameworks to guide the collection and analysis of information. These frameworks must be based on expert knowledge from conventional science and traditional knowledge systems and must produce output that will be accessible and relevant to policy and management needs. The environmental policy indicators approach—based on the Pressure-State-Response model used by OECD countries and embraced by three recent reports (Adriaanse 1993; EPA 1994; Hammond et al. 1994)—represents an important step forward in making scientific data socially and policy relevant.

This framework has yet to be applied, however, to land-based resources in order to define sets of georeferenced ecosystem indicators at the regional ecosystem scale. Hammond et al. (1994) suggest such indicators should represent the life support capabilities of ecosystems or habitats. The practicality of ecosystem indicators for decision making in policy and management contexts is further emphasized by Hammond et al. who indicate that though biodiversity can potentially be measured at the species level, biodiversity preservation must be focused on more encompassing levels such as ecosystems or habitats. They argue that these levels correspond more closely to the scale of geographic units that are practical for assigning administrative responsibility.

Such a policy-indicators framework should incorporate the concerns, concepts, and data most relevant to assessing the development needs at the district or national level (although it potentially can be applied on any level) and should be based on the measurement of the human pressures on, the status of, and the risk to the natural resource base. The specific indicator outputs should include measures of the condition of the ecological subsystem of the socioecological system and the nature of the interactions between the ecological and human subsystems. The multidisciplinary working group developing this approach (ISD/RIVM/WRI 1995) concludes that the framework and its output ideally should provide the following: (1) a reliable way of assessing resource condition; (2) analytical procedures to define factors responsible for resource degradation; (3) policy alternatives to prevent further degradation or accomplish restoration; and (4) analytical procedures to determine whether the policies accomplish their stated goals.

The framework meeting these criteria can be described in reference to the well-established Pressure-State-Response approach and the conventional systems approach of cause and effect used in the EPA's "ecorisk" model. Figure 2.2 illustrates how various conceptual pieces of the framework fit and build on established and accepted methodology. The operationalizing of this framework can be shown in terms of real georeferenced data (figure 2.3). It can be seen from figure 2.3 that a set of descriptor states can be described and measured to produce an aggregate measure (an indicator value) of pressure and of condition that together yield a vulnerability indicator. The values of these indicators are calculated on the basis of georeferenced or spatial data (Wilcox, Smallwood, and Kahn 1999). The functional relationships are modeled on a theoretical understanding of the dynamics of ecosystem and biodiversity change because of human and natural stresses and are calibrated by using specific data on the region under study. The effects of population change on the reduction in forest area can be quantified, for example, using general models, tested for reliability, and calibrated to country or district based on data from the region. These estimates of forest area loss can then be scaled to various measures of biotic impoverishment or changes in ecological integrity. One such approach is to employ a fragmentation model that takes into account the effects of habitat fragmentation on the persistence of key vertebrate species and other biophysical effects of change in land cover. Although this indicator methodology aims primarily at measuring pressures, conditions, and vulnerability of regional landscapes of predominantly natural habitat (Gallopín 1994), it also provides a conceptual description of a general indicator approach that is applicable to predominantly agricultural landscapes or regional ecosystems.

A critically important feature of this approach is that it considers ecosystem or land-unit condition in terms of a baseline state, described by both scientific measures and human values. It is taken as a basic

Bruce A. Wilcox

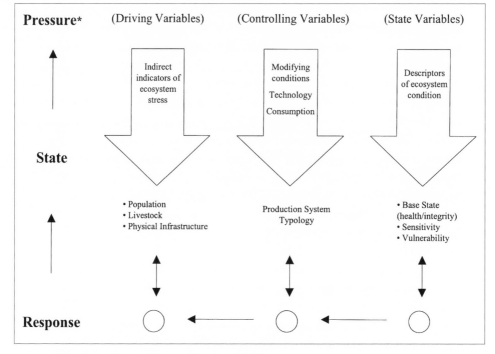

Figure 2.2. Ecosystem Indicators Model

In the middle of the figure, human pressures such as human population density, stocking rates, and infrastructure (for which statistics frequently are available on a district level) are seen as the driving variable and also as descriptors of State (of the system at a given time). However, State is also given by the production and land-use system, as well as by descriptors of the condition of the system (a particular land unit) in terms of its ecological and biodiversity attributes. The causal chain is seen as consisting of human pressures, linked by the modifying influences of production-system type (which may exacerbate or ameliorate the pressures) to the condition of the land unit's living resource base.

Note: * = Overall pressure in terms of the effect on ecosystem functioning and biodiversity within a specified land unit.

premise of the framework that description of a healthy or desirable state is a value judgment that must be made by the local resource users, perhaps in negotiation with others who may be affected. The role of science is to assist in defining the human activities that can be tolerated by the system (and to what degree) and those that cannot, in order to

Geographic-Based Assessment

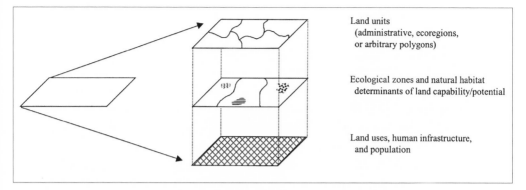

Figure 2.3. Georeferenced Data Layers for Ecosystem-Based Indicators.

achieve or maintain a given desired base state. What is a desirable state and how resources can be managed to achieve different states are questions largely dependent on the sociocultural and ecological conditions specific to a land unit and, thus, on an area's indigenous cultural and biological resources.

Conclusion

As noted by Pichón and Uquillas (this volume), rural poverty and environmental degradation are endemic throughout the LAC region. The juxtaposition of these two problems represents a central dilemma for sustainable development: the need to raise income levels (which necessarily means higher levels of production and consumption) while conserving, if not enhancing, the land resource base. Yet, the possibility of resolving this dilemma is perhaps greater in the rural than in the urban-industrial sector where production modes and values neither reflect ecological realities nor are primarily concerned with renewable resources. Indigenous resources, in terms of both sociocultural attributes (including knowledge systems) and biophysical attributes, provide a foundation for assessing needs, opportunity, and risk relating to rural regions. Such assessment is greatly facilitated by GIS technology and georeferenced data.

The two indicator approaches described in this chapter provide complementary outputs in this regard. The first approach, used as a basis of assessing the utility values associated with land units such as ecoregions, provides a means of determining the values and opportunities that would be threatened should environmental degradation occur or, more commonly, continue. The policy indicator approach can be applied to the problem of assessing the status of land units in terms of ecological integrity or degradation and the likelihood of a change therein, based on current pressures. Although a more detailed explanation is not possible here, the combination of the two sets of data and approaches can potentially provide a more rational basis for development assistance and policy formation. The result should be greater harmonization between the perceptions, needs, and production modes of the rural and urban-industrial sectors. This should in turn result in increased economic development opportunities for rural smallholders, consistent with sustainable resource management objectives at the regional ecosystem scale. The overall outcome will be the enhancement and conservation of globally as well as locally significant indigenous resources.

Conventional thinking about development, which generally treats indigenous resources as having minimal relevance in the global industrial economy, must be replaced with new thinking that views such resources as part of the existing capital upon which development can be based. This change, along with the information upon which rational development planning decisions can be made, can be assisted through the use of a systems perspective applied within a multidisciplinary context to georeferenced data of sociocultural and biophysical relevance. The resulting conceptual frameworks, analytical methodologies, and the information generated should, it is hoped, provide some guideposts along the path to sustainable development.

3

Combining Indigenous and Scientific Knowledge to Improve Agriculture and Natural Resource Management in Latin America

Billie R. DeWalt

Recent years have seen an increase in polemical debate about how well agricultural science and technology have performed in developing solutions to the parallel problems of how to increase agricultural production and how to alleviate world hunger. Individuals who argue that agricultural science has been effective in increasing food production—and thus staving off Malthusian predictions of mass starvation—take the position that further applications of science are needed to meet the demands of continuing population growth. On the other hand, especially among humanists and social scientists, there is increased questioning of the effectiveness of agricultural science and technology in dealing with these problems, because their application has not led to socially just or ecologically sustainable societies. Although worldwide food production per capita is greater than ever, much of this food goes to those with an already affluent diet, while masses of people are still starving. There is growing recognition that the technologies used for increased production are not sustainable and, in many cases, are environmentally damaging (Brown 1989; Commoner 1971; Hightower 1973).

Construction—or in many cases, reconstruction—of a more sustainable and socially just agriculture has led many individuals to argue that greater attention should be paid to indigenous knowledge systems (Brokensha, Warren, and Werner 1980; Richards 1985; Thrupp 1989; Warren 1987, 1991a, 1991b). Their arguments are based on the need to (1) create more appropriate and environmentally friendly technologies; (2) empower people like farmers to have greater control over their own destinies; and (3) create technologies that will have more just socioeconomic implications. This literature has often been ignored by agricultural scientists because of the sometimes missionary fervor with which its proponents preach the virtues of indigenous knowledge systems.[1] Scientists are blamed for the ecological and inequality problems that exist, and the implication is made that all we need do is learn the local knowledge systems of farmers and we will have many of the answers to development ills. Understandably, agricultural scientists are wary of such perspectives.

In a recent paper (DeWalt 1994), I presented a framework for comparing indigenous knowledge and scientific knowledge systems. I would now like to revisit that framework and use it in an analysis of several development efforts in Latin America with which I have been associated. My purpose in this chapter will be to establish a framework for creating more effective and creative interactions between indigenous and scientific knowledge systems. First, I will discuss the strengths and weaknesses of each system. Then I would like to examine three cases of agricultural or natural resource management in Latin America to illustrate the strengths and the limitations of indigenous knowledge systems. I will draw on these examples to indicate in which situations we should look for guidance and ideas from indigenous knowledge systems. The chapter closes with a discussion of how the scientific knowledge system must change, if it is to work better with local knowledge systems to improve agricultural and NRM systems.

Traditional Scientific Knowledge Systems

A variety of critics ranging from postmodernists to political economists have critiqued "traditional" scientific knowledge systems on a number of grounds. These include the value assumptions from which these systems depart and the ways in which the scientific knowledge system fits with and reinforces the interests of the powerful in our social, economic, and political systems (Feyerabend 1975; Lyotard 1985). These critiques, as applied to scientific knowledge systems relating to agriculture and natural resources, are summarized on the left-hand side of table 3.1.

Table 3.1. Characteristics of Current Knowledge Systems Applied to Agriculture and Natural Resource Management

Traditional Scientific Knowledge Systems	Traditional Indigenous Knowledge Systems
Means Used to Study Phenomena	
Specialized, partial	General, holistic
Based on experimentation	Based on observation
Resource Utilization Characteristics	
Dependent on external resources	Dependent on local resources
High input	Low input
Land intensive	Land extensive
Labor saving	Labor demanding
Market risk	Environmental risk
Complicated technologies	Simple technologies
Specialized adaptive strategies	Diverse adaptive strategies
Global sources	Local sources
Outputs	
Low productivity for energy inputs	Low productivity for labor inputs
Cultural disjunctions	Culturally compatible
Profit goals	Subsistence goals
High potential for degradation	Low potential for degradation
Not sustainable	Sustainable with low population densities
Knowledge Characteristics	
Immutable mobiles	Mutable immobiles

Kloppenburg (1991:530) provides a philosophical basis for these critiques of science by arguing that the basic approach to producing scientific facts is Cartesian reduction, the process of "breaking a problem down into discrete components, analyzing these separate parts in isolation from each other, and then reconstructing the system from the interpretation of the parts." He follows Latour (1986:7–14) who has argued that the goal of science is to produce "immutable mobiles"—information that can be transferred without modification to any spatial or social location. In these views, traditional Western scientific knowledge systems search for knowledge that does not change from one context to another (it is immutable) and that should, therefore, be easily relocatable from the specific circumstances in which it is created to other contexts (it is mobile).

There are several strengths of traditional scientists. They learn an extraordinary amount about limited areas of knowledge. They become astute about the principles or mechanisms by which things work, through the construction of theoretical knowledge. They have a very effective means, the scientific method, by which to approach problems and to engage in explanation. The knowledge that they produce is transferable across time, space, and social setting.[2]

As practiced, these strengths have also created problems for science, however. The reductionism of science often leads to woeful ignorance of the wider context within which the particular phenomena under study occur. One problem is that the selection of phenomena to be studied is determined by the ability to break it down to "researchable pieces." Complex systems and those characterized by myriad interactions are likely to be ignored. A second problem is that scientists often advocate the change of one part of the system without paying attention to the results for the overall system.[3] A third problem is the tendency to focus only on the short term, not looking at the potential long-term implications of a change in technology.

More problematic, perhaps, is that science has created a certain hubris among its practitioners. Because of their isolation, many scientists have lost touch with their ultimate goal. Kloppenburg (1991:530) states: "As Cartesian science is elaborated and institutionalized in laboratories, it loses touch with the local knowledge and everyday experiences." At its worst, this attitude leads to an assertion that research is value-free and that scientists need not be concerned with the ethical, social, or ecological consequences of their research. More benign, but potentially as dangerous, is that scientists denigrate the knowledge and experience of nonscientists.

The applications of science to agriculture and natural resource utilization have led to the characteristics listed in table 3.1. Globally, agricultural systems are increasingly dependent on high inputs of external resources such as fertilizer, machinery, and pesticides. They are increasingly specialized (monoculture is dominant), and oriented to satisfying the needs of a fickle market, creating familiar boom and bust cycles in agricultural products like bananas, coffee, sugar, cacao, cotton, and a host of others. They are labor-saving and land-intensive (see Ruttan and Hayami 1990). Global sourcing of food and agricultural products (see Friedman 1994) has become as commonplace as that occurring in industrialized production. The consequences of these systems that search for profit as the ultimate goal include relatively low productivity for energy inputs (see Pimentel and Pimentel 1979); cultural disjunctions as corporate interests and enterprises replace family and community-based production systems; and potentially dangerous degradation of ecological systems. The high energy inputs and natural resource destruction caused by these systems make it ever more apparent that they are not sustainable.

Traditional Indigenous Knowledge Systems

Partially in reaction to critiques of traditional scientific knowledge systems, a variety of individuals have recommended a return to the study and utilization of indigenous or local knowledge systems, for us to learn how to solve the world's problems (Brokensha, Warren, and Werner 1980; Richards 1985). Vanek (1989:167) notes:

> Some features of indigenous knowledge which give it salient relevance to sustainable development planning are its conformity to high labor and low capital demands; dynamics, having evolved over centuries; locally appropriate nature; cognizance of diversified pro-

Billie R. DeWalt

duction systems; emphasis on survival first and avoidance of risk; rational decision-making; various adaptive strategies for use at times of stress (e.g., drought and famine); ingenious system of inter-cropping; integration with social institutions; and flexibility, with considerable potential entrepreneurial abilities.

As with science, however, we need to recognize the potential strengths and weaknesses of indigenous knowledge systems. Table 3.1 includes a summary of the characteristics of indigenous knowledge systems.

Perhaps the greatest strength, as well as the greatest weakness, of these knowledge systems is that they are local. As Kloppenburg (1991) has pointed out, local knowledge produces "mutable immobiles"—relatively malleable knowledge that is finely tuned to the continually changing circumstances that define a particular locality. There are several comparative advantages enjoyed by local people. They are very astute about their local environment and have accumulated a lot of experience concerning those things that affect their existence. They have a keen awareness of the interconnectedness and ecology of plants, animals, and soils. They are ingenious at making do with the natural and mechanical resources at their disposal.[4] The problem is that indigenous knowledge is perceived as rich in contextual detail but it is immobile, having little utility outside particular places (Kloppenburg 1991:531).

In terms of resource utilization, indigenous knowledge systems are likely to be more dependent on local than on external resources. People usually employ a diversity of adaptive strategies in order to survive. Agriculture is generally low-input, land-extensive (although in parts of China, Indonesia, the Philippines, and some other regions this is not the case), and heavily dependent on family and community labor. Adverse environmental conditions (rainfall, floods, frosts, and so on) pose the main risks for these systems. In terms of outputs, these systems most often have evolved in concert with culturally compatible forms of organization. Although most have relatively low labor productivity (because their main goal is provisioning of the family), they are sustainable at low population densities.[5]

Employing the contrasts delineated thus far between scientific and indigenous knowledge systems, we can now examine several cases from Latin America that elucidate their strengths and weaknesses. These cases demonstrate the complementary and productive relationship that should exist between scientific and indigenous knowledge systems. They will serve as means to develop a framework for integrating these knowledge systems.

Specific Cases

The Langosta in Honduras: The Limits of Indigenous Knowledge Systems

An interesting example of the limits of indigenous knowledge systems comes from southern Honduras, a region where the International Sorghum/Millet Project (INTSORMIL) has been working since 1981. With a team of researchers in this region in 1981 and 1982, I conducted a baseline study that focused especially on farming systems used by small farmers. One of our objectives was to identify, from the farmers' perspective, the most important constraints to production, particularly in relation to sorghum, which is an essential component of their production and dietary systems (see K. DeWalt and B. DeWalt 1989).

Our research indicated that farmers had evolved a sophisticated intercropping system of maize (a crop indigenous to the Americas) and sorghum (a crop indigenous to Africa and India) that fit well with regional climatic conditions and with the subsistence needs of the people (B. DeWalt and K. DeWalt et al. 1982). The system developed by local farmers combined a maize variety with a very short growing season (planted at a time during the cropping cycle that would enable it to mature before an often extended dry period in the middle of the rainy season) and a sorghum with a very long growing season (interplanted with the maize but better able to withstand the dry period). In spite of this refinement in terms of crop varieties, the local farmers were unable to cope with some of the major insect pests in the region. We (B. DeWalt and K. DeWalt et. al 1982:42) reported the following:

> The major insect pest mentioned by farmers in Pespire was the *lan-gosta* (locusts). The *langosta* seems to come in waves and to leave most fields untouched while wreaking major destruction on other

fields. Farmers report that they have no way of knowing when and where an outbreak of *langosta* will occur. Consequently, few farmers take any precautionary methods against the insects, which generally do most of their damage when the plants are quite small.

Farmers and extension agents in the region could tell us very little about the insect and its life cycle although the problem had existed for a long time. José del Valle (a famous writer of the early nineteenth century) published a pamphlet in 1804 entitled *Instrucción sobre la plaga de langosta: Medios de exterminala, o de disminuir sus efectos, y de precaber la escasez de comestibles* (Instruction on the langosta problem: Means to exterminate it, or diminish its effects, and to prevent the scarcity of foodstuffs). The pamphlet reported that the langosta caused substantial damage to crops, which increased the threat of hunger. It indicated some understanding of the ecology of the pest, reporting that the female lays her eggs in untilled *(inculto)* areas. The integrated pest management techniques recommended by del Valle (1804:3) included trying to destroy the eggs by plowing, burning fields, or letting pigs loose to root them out. Neither indigenous knowledge among farmers nor scientific knowledge among agricultural researchers and extensionists had resulted in any progress in controlling the langosta over the nearly two centuries since del Valle's description.

INTSORMIL natural scientists developed an interest in the problem because of the damage caused by the langosta. Their research has found that the langosta is in fact a complex of noctuids, which includes at least four distinct species—*Spodoptera frugiperda* (J. E. Smith), *Metaponpneumata rogenhoferi* (Moschler), *Spodoptera latifascia* (Walker), and *Mocis latipes* (Guenee). On-farm research in 1988 and 1989 determined that these various pests differentially affect important cultigens like maize and sorghum at different times in the cultivation cycle (Portillo et al. 1991). *S. frugiperda, M. rogenhoferi* and *S. latifascia* showed a distinct preference for maize and *M. latipes* infested both maize and sorghum equally. Depending on rainfall and other climatic conditions, the pests might shift their feeding habits between weeds and the cultivated crops. Insecticide spraying had little effect on the pest complex. Portillo et al. (1991:295) concluded:

The occurrence of the lepidopterous pest complex *(langosta)* on sorghum and maize creates the biological illusion that the component species coexist conjointly. Actually, the complex is the product of a fine-grained mosaic of different micro-habitats, each supporting a well-adapted successful species.

Because of the geographical and biological diversity involved by the complex, controlling the langosta is a formidable challenge. Although many farmers spray their crops after the langosta has arrived, this has no effect on the core population of at least two of the species, nor does it reduce their subsequent generations since these species are unable to complete or appear to have difficulty in completing their life cycle on maize or sorghum.

The authors indicated that a much greater understanding of the ecology of each of the pests is needed before effective control strategies can be developed. They suggested that an array of integrated pest management practices offers the greatest possibilities for suppressing the complex. Because several of the insects breed primarily on weeds rather than the cultivated crops, chemical controls are likely to be too costly and ineffective.

This case illustrates the limitations of indigenous knowledge systems. Although the langosta were identified as a significant limitation to production as early as the beginning of the nineteenth century (del Valle 1804), farmers have been unable to develop any significant understanding of the problem or ways to control it (immutable mobiles). Although a demonstrated means to mitigate the damage of the langosta complex has yet to be found, careful research by scientists has identified the different pests involved in the complex and suggested the future directions that must be taken for effective control (mutable immobiles).

Tropical Forest Management: The Limits of Scientific Research

Our second case study is from an area where indigenous knowledge systems appear to have considerable potential for contributing to sustainable NRM solutions. Perhaps the most critical agricultural research issue of the next century will be to determine effective, sustainable management systems for the humid tropics of the world. Science and technology have thus far had little success in providing viable solutions for these regions. Intensive resource extraction—like that involved in logging, mining, or petroleum processing (Natural Resources Defense Council 1991)—is clearly destructive for these important ecosystems. Livestock schemes, where the tropical forest is replaced by pasture, and large-scale agroforestry schemes such as the infamous Jari project in Brazil (Fearnside and Rankin 1982) have, to date, also not proved sustainable alternatives.

Monocropping is unlikely to prove a viable option. A good example of the experiences of monocropping comes from the Oriente of Ecuador where naranjilla *(Solanum quitoense)* has proved an economically profitable crop. The juice of this fruit has become extremely popular in the cities of Ecuador, and each week thousands of boxes are trucked out of the upper Amazon region on roads built initially to develop the oil industry. Naranjilla grows naturally in the forests of the region, but because of its commercial value, it has increasingly been planted as a monoculture. After a few short years of monocultural production, the soil and plants become infested with nematodes. The only solution is for the local farmers to abandon their fields and clear another area of primary forest for their plantations. The net effect is a form of shifting cultivation that leads to massive deforestation and destruction of biodiversity. The results of monocropping for pasture, coffee, cacao, pines, and other cultigens in the humid tropics have thus far proved equally unsustainable.

Although little in-depth agricultural research has been done in such regions of the world, it may be argued that the humid tropics do not lend themselves to the kind of reductionist agricultural research that breaks problems down into their constituent parts. The extreme biodiversity of the humid tropics, with low concentrations of plant, insect, or animal life within any delimited area, indicates a greater need for research on ecological systems. It is in such situations that indigenous knowledge systems may be maximally useful as a guide for scientific research.

An alternative to the destructive effects of monoculture is suggested by the research results of Dominique Irvine (1987), who studied the Runa of San José in the Ecuadorian Amazon. The Runa of San José obtain food through gardening, hunting, and fishing. They grow maize, coffee, and cacao as market crops—and manioc, which provides a large part of their consumption needs—in shifting cultivation gardens cleared using slash-and-burn techniques. Much of the effectiveness of Runa subsistence, however, comes from their knowledge of how to manage the succession within this shifting cultivation system. Rather than simply abandoning a garden site to fallow, the Runa engage in management for what Irvine calls "resource enhancement." Such management includes selectively "weeding" naturally occurring pioneer species; protecting or occasionally transplanting desirable fruit, palm, and other trees; and planting trees such as coffee and cacao. The "fallows" also are managed to serve as game attractors to enhance hunting success. The fruit trees that are protected in the fallows serve as food sources for game animals like rodents such as paca and agouti. The result, in comparison with unmanaged fallows, is greater diversity of species and greater economic and subsistence value in the new forest canopy (Irvine 1987: 85–101, 140).

The goal of Irvine's research was not simply to illustrate the wisdom of indigenous techniques of forest management under conditions of low population density. Instead, she wanted to show how their technology was changing in response to growing population density. These are not communities whose traditions limit them to survival only under invariant ecological and social conditions. Irvine (1987:188–89) concludes that:

> My study of succession management suggests that the "seeds" of agricultural intensification are found in the agroforestry cycle. Rather than being bound by the problems of soil management inherent in continuous root crop cultivation, people can supplement these staples and extend the agricultural cycle by augmenting tree crop production. There is considerable variation within San José in the degree and manner of fallow management. I would argue that this variability indicates that intensification is an option for increasing land productivity. . . . [P]opulation pressure resulting from prolonged settlement has encouraged a degree of resource enhancement through succession management.

Unfortunately, the limited agricultural research that is now occurring in the Amazon region of Ecuador is focused mainly on "traditional"

Western agriculture—producing row crops or trees in pure stands. Approaches based on mutable immobiles have not produced positive results. Few efforts are based on working with the immutable mobiles of the resource-enhancement strategies of the local people (see, for example, Redford and Padoch 1992). Instead, their forest management practices are being threatened by public policies that have promoted the expansion of the agricultural frontier, especially the conversion of forest to pasture (Uquillas 1993).

Several years ago, the Bruntland Commission report, which provided the main impetus for the current focus on sustainability, provided comments on the importance of indigenous knowledge systems such as the Runa's. The report (WCED 1987:114) indicated that

> These communities are the repositories of vast accumulations of traditional knowledge and experience. . . . Their disappearance is a loss for the larger society, which could learn a great deal from their traditional skills in sustainably managing very complex ecosystems. It is a terrible irony that as formal development reaches more deeply into rain forests, deserts, and other isolated environments, it tends to destroy the only cultures that have proved to thrive in these environments.

The cultural diversity produced by the human experience is being eroded faster than the biological diversity of the planet. In my estimation, this is cultural wasting—the systematic process whereby the unique social, technological, moral, expressive, and other indigenous knowledge of groups is lost as people become absorbed and incorporated within the world system (see DeWalt 1984:261; DeWalt 1988).

The Seed Drill in Mexico: Indigenous Knowledge Adapting Scientific Research

The final case study illustrates a pattern that probably has thousands of parallels in many parts of the developing world, especially in Latin America where bimodal agriculture is the norm (Johnston and Clark 1982:70–71; Johnston and Kilby 1975). In these situations, the vast majority of agricultural research and extension is oriented toward the "modern" agricultural sector of the *latifundio*. The farmers of the *mini-*

fundio or small farm sector are generally left to their own devices, adopting technology that is more scale-neutral such as insecticides or herbicides and sometimes adapting technology to make it more appropriate to their needs and resources. Farmers in rural Mexico provide a good example of this technological adaptation.

Farmers in the Temascalcingo Valley of the Mexican central highlands in the northwestern corner of the state of Mexico face similar problems to those faced by small landholders in many regions of the world. For the most part, they still use their own labor power or that of animals to work their plots. With small amounts of land, which under the best of conditions provides minimal cash income or subsistence production, they have to employ a diversity of strategies in order to survive. For them, cutting the costs of production is essential.

When I studied an ejido in this valley in the early 1970s, one of the most expensive and time-consuming parts of their maize production system was the planting process (DeWalt 1978). The traditional method of planting in Temascalcingo—as in many other areas of Central and South America—was with the digging stick *(pala)*. Fields were furrowed in crisscross fashion using a team of animals and a plow. Where the furrows crossed, the digging stick was used to make a hole, wherein seeds were dropped and covered using the other end of the digging stick or by kicking dirt over the hole. This labor-intensive method of planting required between ten and eighteen days per hectare, approximately 30 percent of the total labor cost of maize production. As more farmers began to participate in off-farm labor activities, there was a need for a faster method that did not require hired labor.

Local farmers had seen the seed drills pulled by the tractors used by wealthier farmers, and a few had experimented with cup-feed drills pulled by animals. The first method was too expensive, the second unsatisfactory. The farmers then began to experiment. One of the early methods they used in the 1950s was a tube fashioned from maguey leaves and attached to their moldboard plows. While one individual plowed, a second followed behind dropping seeds into the tube. These implements had several disadvantages: they were not very durable and still created a furrow that often left the seed exposed.

In the late 1960s, a local blacksmith addressed the problem by fashioning an inexpensive, dedicated seed drill *(sembradora)*, which (1) had a share that penetrated the soil to a depth of about twenty centimeters; (2) did not have a moldboard that turned the soil; (3) had a metal tube welded to the handle; and (4) sold for about twenty U.S. dollars in 1973. This seed drill allowed two individuals to quickly and efficiently plant a hectare of land in a few hours. Costs of production for a hectare of maize were reduced by almost 25 percent. The seed drill was almost universally adopted in the community I studied. Only a single family continued to plant its maize with a digging stick. The seed drill was adopted much faster than the elements of green revolution technology (improved seeds, fertilizer, pesticides) being promoted by change agents in the area (DeWalt 1978, 1979).

Seed drills—similar to those still used today in and around the Temascalcingo Valley—are depicted on tablets from nearly four thousand years ago in Sumeria (Kramer 1956:62). Science and technology have created many variants of seed drills over the centuries, but a type appropriate to the conditions (that is, relating to labor costs, investment resources, and general economic conditions) of the small farmers in Temascalcingo was not available. Even though the need existed, it was unlikely that modern farm implement manufacturers, agricultural extension agents, or the scientific establishment would ever have worked on devising the appropriate technology. In this case, indigenous technology involved adapting and modifying expensive technology to be more congruent with the local circumstances.

Discussion

The cases described above indicate that we should not rely solely on the findings of agricultural scientists nor on the indigenous knowledge of farmers. We should take advantage of the creativity and innovativeness of both groups. It is important that we see indigenous knowledge systems and scientific knowledge systems as complementary sources of wisdom. In some cases, such as that of the Honduran langosta, a scientific knowledge system with a developed methodology for determining the ecological habits of insects was able to determine the complex of pests causing damage to crops. Such an understanding is the first step in designing appropriate solutions. In the case of management of agriculture in humid tropical regions, science has thus far made little progress. To be sure, ecologists and others have come to an understanding of some of the interactions in such ecosystems, but this knowledge has yet to be applied systematically to designing sustainable agricultural systems. Indigenous knowledge systems, such as those of the Runa, provide some useful guidelines concerning potential future directions of scientific research. In the case of the seed drill in the Temascalcingo Valley, because the scientific and commercial agricultural establishment was unlikely to design and market an appropriate device, farmers and a sympathetic blacksmith took a leading role in adapting and applying technology to their own needs.

We must recognize that both those who use and develop indigenous knowledge systems (mutable immobiles) and those who develop and apply scientific knowledge systems (immutable mobiles) are constrained by the way they have been trained to think and by the contexts in which they live. A substantial amount of literature has focused on how the cultural attributes, traditions, and conservativeness of peasants (see Rogers 1969) and other "traditional people" have impeded development. What is now becoming apparent is that there also are a number of barriers in Western knowledge systems—or what philosophers like Feyerabend (1975) call "belief systems"—that must be overcome.

The first negative aspect of the scientific knowledge system is the blind faith placed in technological determinism or the technological "fix." Throughout this century, there has been a pervasive belief that technology is the ultimate answer to world food-production and poverty problems. This faith in the efficacy of technological determinism is reflected in the way we refer to the "revolutions" created by the advance of technology. The Agricultural Revolution, Industrial Revolution, and the Green Revolution are all seen as hallmarks of human progress, made possible by the advance of technology. Many individuals now look to the coming Biotechnology Revolution as the answer to world food-production problems. By producing more food more efficiently, it is believed, there will be enough for everyone—though not one of the earlier revolutions ever solved the problems of poverty and starvation.[6]

Related to this belief is another: that there exists a "magic bullet" solution that will resolve a myriad of problems. Rather than seeing improvements of NRM systems and agriculture as the result of many
small modifications in complex systems, the magic bullet approach looks
for miracle crops, seeds, or trees. In just the past two decades, eucalyptus,
leucaena, velvet bean, IR-8 rice, tilapia, opaque maize, and a host of other
pretenders have not lived up to their advance billing as potential sav-
iors. The current efforts to breed high lysine maize and sorghums and
to biotechnologically engineer superplants are part of this same syn-
drome.

Scientists are encouraged and rewarded for working on topics that
are on the "frontiers of knowledge," that is, they are not rewarded for tin-
kering with existing technology and practices to make these better or
more productive. Unfortunately, research structures give greater priority to fashionable topics—those that are likely to win individuals praise
within the scientific community—rather than the problems of the poor.
In retrospect, while the award of the 1970 Nobel Peace Prize to Norman
Borlaug brought a focus to the efforts of scientists working in international agriculture, it also reinforced the notion that high-yielding "miracle seeds" (though dependent on irrigation, fertilizers, and pesticides)
were where the rewards were to be garnered.[7]

The politics and structure of biological agricultural research have
actually acted to create some of the problems in international agriculture today. One aspect of the structure that is extraordinarily important is the problem of specialization, both within and across disciplines.
Most scientists today work on basic research issues that have to do
with very specialized topics within fairly specialized disciplines. This
means that scientists are far removed from the ultimate outcome of
their research; they are like the factory worker who puts one part into
the whole, but who may never even recognize the end product.[8]

This specialization—especially of language and topics deemed important—in the discipline creates separate "cultures." The bearers of
the separate cultures find interdisciplinary and multidisciplinary work
difficult, if not impossible. While working as part of INTSORMIL, for
example, I was surprised to find almost as much hostility and back-

biting among agricultural scientists of various subdisciplines as between these scientists and the social scientists (see K. DeWalt and
B. DeWalt 1989).

A final aspect of this specialization problem is that those who create
the technology do not market it. It is usually left to commercial firms
or establishments to take basic research and turn it into a final product.
Their concern is obviously to use the technology to make a profit; social,
moral, ethical, and social justice concerns are not their responsibility.
This same attitude has, to some extent, been adopted by the international agricultural research centers. In part to shield themselves from
criticism, they explicitly state that they are involved in the production
of intermediate goods—germplasm, training, and other expertise—
and that national programs are responsible for the production and dissemination of the final technologies (see DeWalt 1989). Few national
programs have this expertise, which means that commercial firms are
provided with an opportunity to step into the void, and these firms are
unlikely to put social justice or environmental concerns high on their
agendas.

Finally, the reason the traditional scientific knowledge system has
not undergone more scrutiny has to do with the implicit assumption
that SCIENCE (in this case, as exemplified in agricultural research) is
value-free and socially and culturally neutral. In social science and the
humanities today, there are all kinds of intellectual trends—going under
the guise of deconstructionism, relativism, hermeneutics, postmod-
ernism, and so on—sharing the fundamental premise that all knowledge is constructed through the interplay between the observer and
the observed. We need not delve deeply into these trends, but the important message to be derived from these perspectives is that any research (in any discipline) is fundamentally part of the political process.
The grant and contract process, our own education, our participation
in a culture and a nation, all help to determine our research problems.
Inevitably, we serve some masters and not others. As scientists, we need
to engage in a frank self-evaluation of our efforts and to recognize who
we are serving.

Conclusions

Using the local knowledge of problems and solutions as a starting point in agricultural research holds promise for the resolution of some of these difficulties. One excellent example is the work of Thurston (1992), who examined indigenous systems of plant disease management. Others are the studies undertaken by Mathias-Mundy and McCorkle (1989) on ethnoveterinary practices, by Reij (1993) on soil and water conservation, and by Redford and Padoch (1992) on lowland tropical forests. At the same time, we have to recognize that farmers know much less about some aspects of agriculture than scientists. Much about the life stages of insect pests or differences between plant diseases, for example, can only be perceived with microscopes or other scientific instruments (Bentley 1989).

Scientific knowledge systems have the advantage that they can broaden the base of understanding and provide a much greater array of options to farmers. Ultimately, in order to be effective, the results of scientific knowledge systems must be incorporated into indigenous knowledge systems. At their roots, the iterative feedback between farmers and scientists is what farming systems research and development, farmer first, and participatory development perspectives try to accomplish (see Cernea 1991; Chambers 1983; Rhoades and Booth 1982; Richards 1985).[9]

Table 3.2 includes the means, resource utilization strategies, and outputs that should be our goals in creating new, more effective knowledge systems that merge the positive aspects of indigenous and scientific knowledge systems. In terms of means, we should try to achieve the holistic understandings that are characteristic of indigenous knowledge systems. The strengths of observation of these indigenous knowledge systems, however, need to be combined with the experimental method of scientists. We should aim for knowledge that falls somewhere between—or that combines—immutable mobiles and mutable immobiles. Our task is to try to identify "mutable mobiles." From my perspective, mutable mobiles would be defined as contextualized, holistic knowledge that can be adapted and applied to similar phenomena in other circumstances.

Resource utilization strategies should not, of course, completely ex-

Table 3.2. Productively Merging Indigenous and Scientific Knowledge
Systems for Agriculture and Natural Resource Management

Means Used to Study Phenomena
Holistic and general
Mixture of observation and experimentation

Resource Utilization Characteristics
Dependent on local resources with moderate mixture of exotic and external resources
Low input with addition of minimal critical inputs
Land intensive
Labor demanding but not labor onerous
Appropriate technologies
Risk averse (to climate and market)
Flexible adaptive strategies

Outputs
High productivity for labor and energy input
Culturally compatible
Food security and comfortable level of living
Sustainable with high population densities
Regenerative

Knowledge Characteristics
Mutable mobiles

clude external inputs. Pesticides, fertilizers, and machinery may be important, but we should seek ways to identify the minimal critical inputs required to make systems more sustainable. Given the world's growing population, it will be necessary to employ land-intensive strategies, but we also need to search for knowledge and technology that is labor demanding to create employment opportunities. Risk-averse, flexible, adaptive strategies should be attuned to ameliorating the vagaries of weather as well as of markets. The outputs from these agricultural and resource utilization strategies should be culturally compatible, should have food security and a comfortable level of living for producers and consumers as a goal, should achieve high productivity for both labor and energy inputs, and finally, should be both sustainable and (if at all possible) regenerative. That is, we should work to create knowledge that will restore and enhance the properties of ecosystems.

Social scientists must recognize that the knowledge we create or report is often very particularist and only tangentially transferable to understanding other systems. In our work, we can facilitate the discovery and transfer of mutable mobiles from one local system to another. Although social scientists have been only peripherally involved, a good illustration of this transfer comes from the Amazon region of Ecuador where an Ecuadorian indigenous federation requested technical assistance from a Panamanian indigenous federation (the Kuna) for designing an NRM plan in their territories. The Ecuadorian federation also sent a group of trainees to the Peruvian Amazon to learn about forest management from the Yanesha Amerindians (Uquillas 1993).

Within the United States, a recent report by the National Research Council (NRC) indicates another positive example of potential change. The NRC (1989:247) notes that "Farmers and other innovators often develop, through their own creativity, new approaches to solving common farming problems." Because of the increasing concerns about rising input costs, damage to the resource base, and the potential health hazards of what has now become "traditional" U.S. agriculture, there is a growing interest in alternative agriculture—or, more correctly, alternative agricultures. Rather than ask agricultural researchers to provide guidelines for these alternative agricultures, the NRC panel undertook case studies of fourteen farms that were being efficiently managed. In other words, in seeking alternatives to the status quo in U.S. agriculture, the NRC sought out local knowledge to provide guidelines concerning potentially productive new directions (NRC 1989). Kloppenburg (1991:523) reports that the NRC "had little choice but to seek out farmers who had themselves developed alternative practices since the agricultural science establishment had virtually nothing to offer." Yet, as the NRC report itself makes clear, these farms were being operated using a mix of alternative and conventional practices. Like the people in the Temascalcingo Valley, American farmers have combined existing science and technology with their own experience and experiments.

The key for those who use scientific knowledge systems is the need to be open to input from indigenous knowledge systems. As the rapid pace of Westernization attests, practitioners of indigenous knowledge systems have rapidly adopted (many would say, too rapidly) the results of scientific systems. Indigenous and scientific knowledge systems can inform and stimulate one another and, in this way, productively merge to create mutable mobiles, in the search for knowledge and practices that are more sustainable.

PART III

Case Studies

4

Organizing for Change—Organizing for Modernization?

CAMPESINO FEDERATIONS, SOCIAL ENTERPRISE,
AND TECHNICAL CHANGE IN ANDEAN
AND AMAZONIAN RESOURCE MANAGEMENT

Anthony J. Bebbington

Intensify or Die?

"Intensify or die" might be a short and to-the-point development challenge for much of Andean Latin America. In a context of cultural change, economic decline, and mine closures it seems evident that unless there is a significant intensification of livelihoods in the region a combined process of land subdivision, Geertzian involution, out-migration,

This chapter is based on several bodies of research: an ODA-funded ODI-ISNAR study of farmer organizations and agricultural research in Bolivia; an Inter-American Foundation review of indigenous federations in Ecuador; and work for Fundagro-Ecuador, also supported by IAF. Several of the studies did not focus solely on the agricultural and NRM work of these federations, nor did they all use the same methods of enquiry. They stretch over several years, beginning in 1988. I am responsible for the argument of the chapter, but a number of these works were conducted in collaboration with many others, in particular, Hernán Carrasco, Teresa Domingo, Adalberto Kopp, Lourdes Peralvó, Javier Quisbert, Galo Ramón, Victor-Hugo Torres, German Trujillo, and Jorge Trujillo. An earlier version of this chapter was published in *World Development* 24.7 (1996):1161–77 under the title "Organizations and Intensifications: Campesino Federations, Rural Livelihoods and Agricultural Technology in the Andes and Amazonia." The editors are grateful to Elsevier Science Limited, The Boulevard, Langford Lane, Kidlington, OX5 1GB, England, for granting permission to reproduce parts of that article herein.

and continued resource degradation will leave large parts of the Andes (especially the altiplano areas) as little more than labor reserves—a situation not so different from some of the predictions deriving from Alain De Janvry's 1981 interpretation of agrarian and rural change in the region fifteen years ago. This chapter, without being quite so dramatic in its language as De Janvry's, can be read in such a way as to point us toward a similar conclusion.

The relative weakness of intensification processes in the region suggests that neither indigenous patterns of technical innovation nor introduced innovations from modern science have been sufficient—either by themselves or in partnership—to trigger this intensification. Indeed, not only will intensification require support for technical change, it also will depend on the rural poor (henceforth "campesinos") having improved access to existing and new product and input markets through relationships that will allow the wealth deriving from natural resource–based and agricultural activities to be captured and reinvested in the Andes. This improved and renegotiated market access is critical for the creation of incentives to the sustainable intensification of natural resource use and of new income and employment opportunities. Technological innovation alone, however environmentally sound and grounded in traditional practices, will not achieve this.[1]

Notwithstanding the bleak picture that can be painted of the Andes, progress in Qiwi-Qiwi near Potosí, Bolivia, or in Salinas in highland Ecuador shows not only that sustainable intensification can occur but also that incentives can be such that rural people will invest in land and reverse resource degradation. The question is how this intensification might occur more widely. This "how" question is closely linked to a "who" question. Although we know that getting the policy environment right and removing perverse incentives is critically important (see Southgate 1992), we also know this is not sufficient (Uquillas and Pichón 1995). Experiences in Potosí and Salinas suggest that, alongside a mix of traditional and modern resource management techniques, modernized forms of indigenous organization and institutional practices also have an important role to play in intensification processes. These roles include negotiating new market relationships, managing on- and off-farm technologies, negotiating new relationships with other institutions, and so on.

Organizing for Change—Organizing for Modernization?

The modernized indigenous institutions and their actual and potential roles in natural resource–based livelihood intensification are the theme of this chapter. We begin by outlining several reasons for looking at these organizations, rather than concentrating on NGOs and public organizations as institutional options for fostering rural intensification. We elaborate on what constitutes these modernized indigenous institutions—their structures, their qualities, their origins. We then look more closely at several experiences of these organizations in natural resource–based intensification and then draw some lessons from these experiences.

The cases reviewed suggest two of the key factors in understanding how, why, and how effectively a campesino federation engages in natural resource and technology development work: (1) the origins and original objectives of the federation and (2) the criteria upon which membership in the federation is determined. From this, we identify three different trajectories whereby federations become involved in technology and natural resource activities: from organized campesino politics, from resource accessing, or from economic and productive activities. We also identify three broad types of roles the federations play in this area of activity: the technology service deliverer, the natural resource management–based social enterprise, or the natural resource pressure group.

Based on these patterns, our general argument is that, although rural peoples' organizations (RPOs) have different approaches to linking the traditional and the modern in their resource management strategies, with time, the organizations that are "successful" (meaning those that continue to receive the active participation and support of their members) are those that foster technical modernization strategies linked to traditional ideas of local control and that contribute significantly and sustainably to the livelihoods (and above all the incomes) of their members. These significant contributions most frequently occur when the organization is involved in marketing, processing, land tenure, and other activities that improve the market options and income possibilities of their members. By implication, the significant technical and natural resource work is linked to these off-farm activities. This is most often the case in the NRM-based social enterprises and, to an extent, the natural resource pressure groups.

This leads us to two additional important points: (1) that the primary significance of these organizations in agricultural and livelihood intensification derives from their potential contribution to regional wealth creation and accumulation in a form that creates further employment and income opportunities in the locality; and (2) that in the specific case of natural resource work, these organizations create positive economic incentives to more sustainable forms of land use (and of this, there are still few examples). When these organizations merely play the technology service deliverer role for their members (for instance, offering technical assistance, seed, and so on), they play a useful role in local development processes but ultimately not a very significant one. Only when they confront and reverse the mechanisms through which natural resource–related value is extracted from a region will they or anyone else begin to make an important contribution to intensification.

We still know too little about the potential of these organizations to be natural resource–based social enterprises that contribute to agricultural and livelihood intensification. There are some significant success stories in this regard, but it is still not sufficiently clear under what conditions and through what means other organizations elsewhere might become effective social enterprises and what sort of investment strategy and complementary support are most likely to facilitate this. If our interest is to link technical and administrative modernization with traditional conceptions of local control and resource management, the theme of socially oriented enterprises will be one very important starting point. (With this idea in mind, IIED is set to initiate a program of research into these rural social enterprises.) The argument is based on research in Ecuador and Bolivia involving a number of organizations. In particular, it draws on relatively detailed fieldwork with federations in Ecuador (Bebbington et al. 1993a, 1992) and in Bolivia (Bebbington et al. 1995).

Why Rural Peoples' Organizations?

The question, Who will support these processes of natural resource and livelihood intensification? leads us to consider the possible roles of

RPOs. As the presence of the public sector in the countryside decreases, and as we recognize more explicitly that the public sector would never have been strong enough to work with but a small percentage of campesinos, it has become common currency that, as Uquillas and Pichón (1995) tell us, we need to look for new constellations of institutional support to campesino livelihoods. Such constellations have been designated "pluralistic participation" by the World Bank (Uquillas and Pichón 1995), though other commentators might have used less generous terms. These constellations will involve National Agricultural Research Services (NARS), NGOs, local governments, universities, and RPOs.

Much is rightly said about the potential of improved government-NGO coordination for generating sustainable agricultural technologies that build on indigenous and modern practices. Trends in NGO work and in NGO relations with government in Latin American agricultural development have been reviewed (Bebbington and Thiele et al. 1993b).[2] A number of the limitations of this earlier review are especially important for the present discussion:

(1) While identifying and analyzing some of these new relationships, the study said relatively little about their impact on rural livelihoods and, in particular, on the development of market and regional accumulation opportunities that might facilitate intensification and technical change.

(2) The study pointed out that NGOs are often not sufficiently responsive or accountable to—and can have strained relationships with —rural people and their organizations. However, the review left pending the question of what types of mechanisms might be employed to make the NGOs and their new relationships with government more responsive to rural peoples' concerns.

(3) The study raised the problem of the institutional sustainability of NGOs but did not pursue in depth the alternative institutional forms that might be more sustainable.

Indeed, as these limitations on NGOs become more apparent, some might argue that NGOs are facing emerging crises of legitimacy and sustainability.

The questions remain. Are these new constellations linking NGOs and government going to help create new incentives and support for sustainable intensification of natural resource management? Are they going to be sufficiently accountable to the rural poor and efficient in their use of resources so as to have a tangible and significant impact on poverty and natural resource management? Are they sustainable? These also are important strategic questions for the World Bank as it begins to foster more involvement of professional NGOs in the programs it supports.

In addition to these doubts about the roles that could and should be played by NGOs and the other actors in the new institutional constellations, there are more positive reasons for looking at the potential roles of RPOs:

(1) Although in many practical respects they are weaker than NGOs and government, RPOs may have more potential to become long-term actors in rural Latin America. Indeed, their relative weakness is largely a reflection of comparative resource endowments. NGOs and government look strong because they receive grants. Without these funds, the NGOs would fold (indeed many do) and government would wither (it often does). Yet somehow these popular organizations continue to exist, with few or no resources. They remain managerially weak, but they survive partly because their strength is based on ideas that are in some sense shared at the grassroots—ideas of self-management, defense of rights, and so forth.

(2) They are organizations that link the traditional and the modern more clearly in their everyday practices than external organizations do, and that as a consequence are potentially more attuned and accountable to, and more grounded in, local social processes. At the same time as being modernizing forces, these organizations have been used by their members to pursue claims of access to resources and rights that are grounded in historical tradition. Although these organizations are often based on nontraditional ideas of organization (unions, cooperatives, and so on), much of their internal functioning as organizations is influenced by traditional ideas and practices (see Rivera-Cucicanqui 1990). Thus, running through all their activities is a clear sense of modernizing from a grounding in tradition. That is, their strategies of modernization are grounded in strong and accepted or historically long-standing local ideas of organization, control, and ethnic identity—again suggesting that these organizations have a greater potential to be sustainable, accountable, and more locally adaptive.

(3) Potentially, RPOs are a type of organization that can be legitimately representative of rural peoples' concerns. Even if their internal mechanisms of decision making are not always participatory these organizations, as the organizations of rural populations, are still potentially representative in a way that NGOs and NARS never will be.

None of this is to idealize these organizations. One can quite easily make the same romanticized overestimation of the qualities of "indigenous," "local" organizations as has been done for "indigenous" and "local" technical knowledge (see comments in Uquillas and Pichón 1995). For the record, let it be said from the outset that, just like NGOs and government agencies, RPOs are sometimes corrupt and dishonest, weakly accountable to their members, inefficiently managed, politicized by party concerns, and overenthusiastic about the virtues of modern technology. In many areas, the traditional campesino federations of the 1960s and 1970s are now heavily criticized for just these reasons. Notwithstanding, it remains the case that RPOs have many of the qualities referred to above, and some of them—assisted by NGOs and others—have made significant contributions to natural resource–based intensification. It is on the basis of these observations that we take a look at several RPO experiences with agricultural technology, with the management of natural resources, and with linking social enterprise and natural resources. In particular, we look at their capacity to generate proposals for agriculture and natural resource management that link traditional and modern ideas; to link agricultural technology development and natural resource management to developing new market and livelihood opportunities and thus contribute to natural resource–based processes of microregional development; and to exercise influence on other institutions in NRM and agricultural technology development.

Who Are These RPOs? The Natural Resource and Technology Literature

It is a cliché to say there is a diversity of types of RPOs with some interest in NRM—but it also happens to be true. We are concentrating here on one particular type of RPO—federations of grassroots groups existing at a microregional or regional level. These are not national organizations, but they are large enough to have the potential clout to influence other technology, development, and natural resource–oriented institutions and to negotiate new and existing market relationships.

Not a great deal of literature about the actual or potential work of these organizations in NRM and agricultural development exists. What literature there is about these organizations deals with them primarily as political actors, paying far less attention to their role as economic actors, and far less still to their acting as natural resource managers (for some exceptions see Esman and Uphoff 1984; Carroll 1992; Tendler, Healey, and O'Laughlin 1988). To the extent that the NRM and technology development literature has become interested in the role of RPOs, it has been through two points of entry. Each has generally led to a concern for local grassroots groups rather than regional organizations. These two points of entry have been (1) farmer participatory research (FPR) and (2) common property resources (CPR) literatures.[3]

The FPR and participatory technology development literature has for a long time been more focused on methods for involving farmers in the research process and for supporting their experimentational abilities than focused on the role of farmer organizations in technology development. Most of the interest in groups has been at a community level or in groups created by researchers specifically for this purpose. The typical tendency was to adopt an instrumental approach to the groups —they were seen as a way to make the researchers' technology development process more efficient and more effective. More recently we have seen interest in working with groups at a larger level than the locality, initially with the aim of increasing the efficiency of the technology development process but, more recently, with a view to how technology development can be a part of the wider strategy of these organizations (see Ashby's work in Colombia; her chapter in this volume; and the work of Poats and Romanoff in Ecuador [Romanoff 1993, 1990]).

The CPR literature with its more generic interest in collective action and management has had a far longer-standing interest in rural social organizations. Some of this work has considered the role of federated forms in managing complex resources whose management requires intercommunal coordination (such as irrigation systems). Most of the work has generally dealt with localized organizations that manage particular resources, such as joint forest management groups, water user groups, and so on (see Orstrom 1990). Again, the emphasis has been on this single natural resource–related function, and the implication of the literature has often been that the research was dealing only with single-function organizations.

These two bodies of experience in work with RPOs have increased their influence over the thinking of key international institutions. This influence is manifest in the increased interest of certain CGIAR institutions in the role of local institutions in natural resource management (such as recent initiatives at IFPRI and CIAT). Certain programs within the World Bank also exhibit increased interest in the same question and in the role of farmers' organizations in agricultural research systems, for example current Bank-supported programs in Mali (Collion 1995).

Campesino Federations

Campesino federations are a somewhat different type of organization from those dealt with in the CPR and FPR literatures, particularly in terms of their origins and in their level of interest in natural resources and technology issues. Although there is a long Andean tradition of supracommunal coordination (through the *etnias*, for example), campesino federations, unlike a number of CPR and community organizations, are not traditional organizations. The federations result from changing political and economic concepts and circumstances and have often been linked to new ideas of organization—for example, centers in the Ecuadorian Amazon, cooperatives in areas of colonization such as lowland Bolivia, and unions in Bolivia. They represent new ways of organizing for indigenous people. Although related to external influences, however, these organizations have in many cases been assumed by indigenous people as their own.

Unlike the organizations discussed in the CPR and FPR literatures, these federations very rarely emerged to deal with NRM and technology issues. Rather, most emerged around questions of rights (land, ethnic, religious, linguistic) or economic activities (marketing, processing). They have only subsequently become involved in NRM and technology activities.

In order to understand the dynamics of these federations, there are various ways we can distinguish among them. When thinking about how they address agricultural and natural resource issues and livelihood intensification in general, it is helpful to categorize them according to (1) their function at the time of formation (usually still their principal function) and (2) the criteria upon which people are or become members. (These two broad dimensions of the organizations will reappear as explanatory factors in understanding how they approach technology and natural resource management as well as the impacts of their work.)

Among the federations referred to in this chapter, we can identify four broad initial functions. There are those federations that play a political and representative role (CIDOB, FSUTC-SC); those that play an entirely economic role (El Ceibo, FUNORSAL, UCIG, AOCACH); those that play an economic role within a larger political or representative organization (CORACA-Potosí); and those whose role was originally simple resource accessing, that is, to access an available resource (UO-CACI, UCIG). From these diverse origins, the organizations have all moved into agricultural and natural resource work, but their origins have clearly influenced how they did so and how effective they have been in doing so.

The federations we consider here also have differing criteria of membership, with significant influence on the effectiveness with which the organizations engage in agricultural, natural resource, and social enterprise work. Above all, these criteria influence how widely the benefits of this work are distributed. There are several criteria:

(1) *Automatic membership via community location with no real barriers to entry.* People are members simply because they live in the area of the base organizations that the federation is meant to represent. This type of membership criterion is found in the political organizations based on principles of inclusivity and representativity and in those organizations created to access resources. These organizations often nominally require a contribution from the member organizations but will presume to represent them even if this contribution

is not paid (for example, CORACA and FSUTC-SC as cases of representative organizations and, to an extent, UOCACI as a case of a resource-accessing organization).

(2) *Membership via cooperative affiliation, with barriers to entry.* Base cooperatives (and other base organizations) become members, but this is not automatic. They have to request membership and make some sort of payment. This barrier to entry is more rigorously enforced than in the previous case. This case typifies economic federations of producer cooperatives and associations (for example, El Ceibo, FUNORSAL, AOCACH).

(3) *Individual membership, with barriers to entry.* Individuals (not organizations) affiliate. They must apply, generally meet certain requirements, and pay some sort of dues. This membership regime is found in economic organizations based on prior individualized activities (for example, UNAPEGA).

In all cases where barriers to entry exist, the general rule is that the higher the payment, the more exclusive the organization, and the more rigorous it is in enforcing this exclusivity of access to its benefits. This has clear implications for how widely the organization's benefits can spread and how far they reach down the poverty scale.

Other sources provide more detail on the experiences of these different federations with technology development and natural resource management.[4] Here we present mere vignettes, to give a broad outline of how different organizations have become involved in these domains of activity. Rather than present each case on an individual basis, the vignettes group the cases according to the path each organization took to move toward these activities. We first look at three broad types of trajectory whereby RPOs have moved into technology work, and then three broad ways that they engage in technology and natural resource–based activities.

Origins of Federations and Their NRM Work

Almost all the federations referred to in this chapter—and perhaps in general—engaged in other types of activity prior to establishing their own agricultural technology and NRM programs. Indeed, the nature of the federations' work in these programs is related to the prior evolution of the organization. Three broad trajectories can be identified: from organized campesino politics to technology, from resource accessing to technology, and from economic activity to technology.

From Organized Campesino Politics to Technology

Many of these organizations began activities primarily as political organizations concerned with claims for, and the defense of, rights. In other cases, as in the union movement in Bolivia, they were created as a political interface between government and campesinos but, in due course, became the organizations through which the campesinos exercised pressure on government. Over time, simply engaging in lobbying and political activity has not sustained the interest of members. The difficulty typically faced by these federations has been that, once they are successful in gaining a right for members (access to land, for instance), they lose their momentum and find themselves with little left to offer. Failing to win their right produces the same result. In order to find new ways of sustaining member interest, the organizations move into new activities. In most cases, this has been done within the same organizations. In other cases (as in the Bolivian peasant union movement's creation of the CORACAs), the organizations have created a special arm devoted to economic activity. In either case, the economic/service work is oriented toward the wider objectives of campesino development and organizational strengthening.

Although in large measure the technology work is a means to a political end, in some cases it can be a political activity in itself. This is the case of explicitly indigenous organizations that commit themselves to the revalidation and revitalization of ethnic identity. For instance, work with indigenous technologies is perceived not only as a technical activity but also as a statement concerning the inherent value of indigenous culture and the importance of strengthening ethnic identity. Indeed, this political commitment can sometimes be so strong it leads the federation to impose a technological agenda on its members, who would rather receive support with modern technologies (Bebbington 1993).

That this technical work has an organizational strengthening orientation—and that this occurs within explicitly representative organizations which aim to represent a wide membership with often varied economic interests—tends to lead these types of organization to prefer activities that can deliver some form of service to as wide a membership as possible. Thus, technology work is often not oriented to particular market niches. It is not treated with the rigor necessary to develop a self-financing program and tends to assume a certain clientelistic dimension.

From Resource Accessing to Technology

A number of other federations have emerged in far more opportunistic ways in order to access a particular resource. Then, having accessed that resource, they looked to other areas of activity in order to sustain (and justify) their continuing existence. For instance, UOCACI in Chimborazo, central Ecuador, was created primarily to access electricity supplies for some thirty different communities and, indeed, continued to administer aspects of the maintenance of the rural electricity system in that area. Having been successful in this, UOCACI sought other activities to sustain the organization. As it was an organization based neither on a long-standing tradition nor on member contributions, it had to seek external financing, and it did so successfully.[5] One of the key activities it won financing for was a technical assistance and seed and input delivery program to its members. The program was administered by UOCACI's team of *promotores* and worked on a form of farmer-to-farmer extension model. It delivered small quantities of inputs to some of its members, because in practice it lacked sufficient resources to give all members support.

In other cases, the initial resources accessed by the organizations—and thus the resources that in practice gave rise to the creation of the organizations—have been specifically earmarked for technology and natural resource work. In these cases, the organizations have in many respects been a creation of the external agencies giving the resources and have been greatly influenced by the concerns of those agencies. In Ecuador, a number of federations were "induced" by rural development projects with the idea that the federation structure would help the project interface with campesinos and also ultimately take on responsibility for project management. For instance, UCIG was in many respects created by the local rural development project and then consolidated by an NGO program for seed multiplication, input supply, and technical assistance in potato production. UNAPEGA in Bolivia grew directly out of an NGO livestock credit program and was created by the NGO to continue managing that program.

Organizations that emerge in this way tend to engage in service delivery work. In large measure, they maintain the logic existing at the moment they were formed, namely, to access a resource and then distribute it to their members. Once an organization has commenced with "free" services and resource distribution, a culture of the "gift" is created within the organization, and members come to expect the organization to continue delivering gifts. For leaders to change this culture and begin to operate on the basis of cost recovery and self-financing becomes difficult (though not impossible) and can become personally costly in terms of leaders' local legitimacy.

From Economic Activity to Technology

The third trajectory is that from a natural resource–related economic activity into direct natural resource and technology support work. This is the least common, and there are fewer such organizations. In this case, the organization sees the need for technical assistance for its members that will maintain and improve the quality of their products so they can keep their market share and, indeed, compete more effectively in the market, access to which the organization negotiates for its members. For this reason, FUNORSAL moved into animal health work in order to increase the volume and quality of the milk being turned into cheese. Similarly, El Ceibo introduced rust-resistant varieties of cacao when an outbreak of rust threatened to damage local production. Later, El Ceibo with the assistance of NGOs and donors opened an organic European cacao market for its members and then began to develop organic production techniques suitable for the Alto Beni of Bolivia. It also gave its members training in these practices.

In this way, federations become the means through which market messages are conveyed to producers and technology development processes are made available. A federation can assume this role precisely because it is directly involved in market operations. This market "discipline" also leads these federations to be more directed and selective in the technical activities in which they engage, that is, they engage only in those for which there is a market pay-off.[6]

The Nature of Federations' NRM Work

Campesino federations have become involved in technology development and NRM work in a variety of ways. However, if we look at their programs, we can see perhaps three main forms of engagement: (1) the technology service deliverer, (2) the NRM-based social enterprise, and (3) the natural resource pressure group. The three roles are not mutually exclusive, though each organization tends to concentrate more in one area than others. Of the three roles, although the first is the most common, we argue that the second and the third are the most important for the sustainable intensification of natural resource–based livelihoods. Why? Because the third is aimed at securing the legal and institutional context within which these livelihoods can be composed, and the second is aimed at fashioning a local economic environment in which there is a stronger possibility that these livelihoods can be created and will be competitively viable. The effectiveness of each of these two roles depends on the other being either fulfilled or unnecessary (for example, where land and natural resource rights are not under threat).

The Technology Service Deliverer

In the role of technology service deliverer, the federation in large measure mimics public and nongovernmental agricultural development programs. That is, it establishes a program wherein the federation hires one or more technical professionals. These professionals then train campesino paratechnicians to give technical assistance in certain areas of agricultural production. A technical assistance program is mounted, which is usually tied to an input delivery program that generally works with seeds, agrochemicals, and animal breeding stock (UNAPEGA, for example). These are usually given as a loan-in-kind at a below-market rate of interest. Work with seed systems, for instance, usually attempts to access improved seed varieties from research institutions and then to distribute this seed through mechanisms whereby the federations co-ordinate community- and family-level multiplication efforts.

This is perhaps the most commonly seen form of engagement with technology development because it is easy. This approach merely reflects federations' adoption of a model developed by other agencies and involves relatively little innovative thinking. It is a model that a number of donor agencies have become used to working with and, thus, are happy to continue to support.[7] The high frequency of this approach is also partly because it is the simplest strategy to pursue, as long as the organization has external funding. The logic of the strategy revolves around the steady distribution of a subsidy received from outside (that is, the donor grant) to the members of the organization (the same logic as that underlying most NGO and state programs). This, in turn, is attractive to the leadership of the organization because it lends itself to a modest form of patronage that can sustain base support for the leadership. As long as the leader can continue accessing grants for seeds, fertilizers, agrochemicals, and so on, then he or she (usually he) retains legitimacy and support.

The NRM-Based Social Enterprise

In the NRM-based enterprise model, the federation assumes a role that is primarily economic (a role that can generally not be played at a family or community level) rather than service delivery or political. This role generally involves product transformation or marketing activities. Among the federations considered in this chapter, the roles include the transformation of cacao beans into powder and chocolate (El Ceibo), milk into cheese and wool into textiles (FUNORSAL), and the marketing of food crops such as potato (CORACA, UCIG). In general, members sell products to the federation, which is responsible for subsequent processing, sales, and all the administration to manage the income that derives from this off-farm activity. This income is then divided among the members, the costs of and investment in the organization's own development, and the costs of some related services.

These activities are generally linked to services. Thus, El Ceibo has a cacao technical assistance program that has introduced improved cacao, worked on blight control, and implemented systems of organic production for organic markets accessed by El Ceibo in conjunction with international fair trade NGOs. FUNORSAL has an animal health program. These activities were initially supported by donor grants and then progressively by the income generated from the economic activities. FUNORSAL's dairy operations and some of El Ceibo's cacao operations are now largely self-financing.

The economic role is where cooperatives tend to be more active, and indeed, FUNORSAL and El Ceibo are both federations of cooperatives. Nonetheless, this is not a role assumed only by cooperatives. Federations of communities (such as UCIG and CORACA) also engage in this role. The ways they come to this natural resource–based social enterprise role tend to differ. The federations of cooperatives come to it having been previously engaged in marketing and so on, whereas the federations of communities have usually been engaged in organized politics or resource-accessing activities and have subsequently seen the need to make a more significant economic contribution to their members' livelihoods.

The Natural Resource Pressure Group

This third, rather more overtly political, role involves federations in exercising voice and pressure on natural resource issues. In general, these demands relate to questions of control and access. Demands are primarily directed at government and usually invoke the idea of rights, demanding that government recognize these rights and protect them against other institutions. Thus, for example, federations engage in lobbying on land tenure issues, forest protection, and territorial issues. Although not all pressure is directly related to NRM issues, such lobbying aims to defend rights that will have a critical influence on how natural resources and technology will subsequently be managed. The lobbying is often linked to discourses on resource management. For instance, CIDOB (the confederation of indigenous peoples' organizations of the Amazonian, Chaco, and other east-northeastern lowland regions of Bolivia) sees an explicit link between its work on the green labeling of timber products and its work in demanding territorial rights for its member organizations. They argue that, if they are able to demonstrate that CIDOB and its members can manage fragile lands sustainably and profitably through programs such as green labeling, the government will more likely recognize those rights, and the CIDOB members will defend them against those aiming to violate them and extract timber unsustainably. This natural resource pressure group role has been most apparent in recent years among federations of indigenous peoples in the humid tropical areas of Andean and Amazonian Latin America. Although some highland federations continue to argue for enhanced tenurial rights, the strength of their cases is less fervent or sustained than in the lowlands.

Some Lessons to Be Drawn

From the cases described we can extract a range of lessons about the work of these federations. We concentrate discussion here on four main themes that build on topics introduced in the chapter by Pichón and Uquillas (this volume). The themes are: the approaches to traditional technology and forms of organization in the strategies of these federations; the socioeconomic impact of their technological activities; their sustainability and viability as organizations; and their potential role in a "pluralistic participation" approach to rural development.

Modernizing from Tradition: Approaches to Technology and Organization

A number of the organizations referred to in this chapter have tried to work with indigenous technologies and have promoted these among their members. They have worked on questions of forest management, minor forest products, native crop varieties, and traditional forms of pest control and fertilization. This has been particularly so in organizations with a strong ethnic identity and a strong commitment to, and engagement in, indigenous politics. In practice, though, their members have demanded different types of support. They argue that they already know much of this indigenous technology and that it is no longer viable within contemporary agroecological and market conditions.

This is one reason a majority of the federations have prioritized work with modern technology. On the basis of local knowledge and some trials, they have adapted this modern technology to local possibilities and conditions. In comparison to the more complex array of inputs promoted by research institutions, the federations recommend only the basic components, such as new varieties, fertilizers, and agrochemicals. In scaling down these modernizing technologies, the federations become vehicles for linking members to sources of nonlocal and nontraditional inputs, which, in some cases, would otherwise be unavailable to members and, in other cases, are less expensive to acquire through the federation than through alternative channels.

That they help to introduce these modern, new practices is not especially surprising. The federations' legitimacy vis-à-vis their members often depends on their ability to deliver something new (or subsidized), so they are drawn toward new technology. Also, the leaders tend to come from a more modernized sector of the local community, so they are generally predisposed as individuals to work with new techniques. Nonetheless, this is still in some sense a modernization grounded in local knowledge, at two levels: the knowledge that farmers already have, and the knowledge that the federations' leaders and campesino paratechnicians already have. This is the knowledge that allows an adaptation of new technologies to local conditions.

At the same time, the federation can become the means through which new knowledge of modern market possibilities is introduced into local practice and through which "traditional" practices can be adapted to those new markets and made profitable. Thus El Ceibo, through links with European agencies, gained knowledge of organic cacao markets and disseminated it to local technology development processes to produce an organic cacao. Organic coffee–producing federations in Mexico offer a similar example.

These federations represent new, indeed modernized, forms of organizing for rural communities. Although they do not necessarily seek to supplant traditional organizations, they constitute an organizational adaptation for new economic, technological, and political contexts. Potentially, they allow access to markets, resources, and power, and they facilitate forms of locally controlled technology development to which traditional organizational forms are less suited. On the other hand, this "newness" of organizational form can easily engender difficulties. For instance, it can lead to problems of accountability within the organizations, because there are no processes for exercising accountability that are entirely appropriate to the new forms.

These are still organizations grounded in traditional concepts and claims, however. They call for local control of natural resources, economy, and politics and aim to enhance that stand. They endorse a strengthened, changed relationship between local populations and the land. Notwithstanding very real problems of internal accountability, they combine new mechanisms of accountability with the more traditional processes and mechanisms of face-to-face relations, *chisme* (gossip, informal social control), and assembly-based discussions and elections.

There is one other important and interesting way these organizations represent a continuity with traditional resource management practices. As numerous cultural-ecological and anthropological studies have demonstrated, traditional organizations are often structured to manage local resources to particular ends. There is a synergistic relationship between technology and organization that is neither deterministic nor pure happenstance. Similarly, these new federations have synergistic structural relationships with new technologies, in that their potential for centralizing and coordinating elements of resource management allows an engagement with new processing and marketing technologies. Their forms allow an engagement with the "institutional technology" through which much new production technology is generated in the modern world. That is, they allow a more effective and assertive engagement with agricultural research institutions, universities, NGOs, and so on. Thus, as the technological and institutional context of resource management in these fragile lands changes, these federations represent one organizational response that will facilitate local engagement within this new technological and institutional context.

The Socioeconomic Impact of Technological Activities

The impact of the work of these federations can be assessed in a number of ways. The following discussion will focus on three: the number of campesinos assisted, the depth of impact on the campesino economy, and the social distribution of benefits. The number of campesinos influenced by the work is often significant. FUNORSAL has twenty-three grassroots organizations as members; El Ceibo and UOCACI, over thirty each. UNAPEGA has about five thousand members. CORACA claims some thirty-five thousand members, though in practice it works with far fewer.

The quality of impact on these members varies greatly. In the service delivery organizations, the impact is short-term and not sustained, involving a short-term injection of technological capital into the members' family economy. This may be relatively insignificant (the odd sack of improved potato seed) or more significant (the loan of improved animal breeding stock). Even in the less frequent cases where it is significant, the impact on the agrarian economy is limited and short-term. The one significant exception to this general pattern is when the federation offers a service that will improve the beneficiaries' control over resources, for instance, a service that helps them secure tenure, as in the case of the cattle development work of the Shuar Federation in Ecuador (though this also had other less positive impacts on the environment).

Some of the most significant impacts are apparent when the federation offers access to markets or product transformation options and follows these with technical assistance that supports market access. In this case, the impact is recurrent, and, rather than offering a short-term and small capital input to the members' family economy, the federation offers a new, sustained relationship between families and the market. This in turn facilitates a more sustained and progressive capital accumulation, in some cases linked to employment generation in the region. For example, FUNORSAL's activities in product transformation have created some three hundred new jobs in Salinas.

The second type of sustained impact is confined to those organizations we called NRM-based enterprises. There also is a link, however, between the type of impact and the organization's dynamics and membership rules. Organizations that become involved in technological activities, in order to strengthen their presence at a base level or in order to recruit new members, tend to prefer to spread the benefits they offer as widely as possible. They prefer to deliver small amounts of one-off (or only occasional) benefits to many families. These organizations—such as the *sindicatos* in Bolivia or some federations in Ecuador, based on a principle of automatic membership granted on the basis of community residence—tend to spread benefits more widely and thinly in order to satisfy the demands of their many members.

Conversely, those organizations that have barriers to entry (where affiliation requires the campesino to pay a significant membership fee and to meet other requirements) limit benefit delivery to those who have met these requirements. Indeed, the benefits delivered by the organization are the incentive to become a member, and the organization will therefore tend to limit spillover effects to nonmembers. It is precisely the NRM-based enterprises that tend to have this sort of membership rule and thus have a more significant impact, but among a more limited membership.

The issue of "barriers to entry" leads us to the social distribution of the benefits deriving from the NR work of these organizations. Evidently, those organizations with higher barriers to entry (that is, the generally more effective organizations) also are more likely to exclude the poorest from their membership, simply because the poorest cannot meet membership requirements. This very exclusivity favors success, because it allows the federations to reduce the complexity of their membership and thus the diversity of the demands made on the organization. The diversity of demands on CORACA, from farmers in five different production systems contexts, is far greater than the diversity of demands on El Ceibo, primarily linked to different elements of cacao production and marketing. In many respects, El Ceibo's task is far easier than CORACA's. In addition, those organizations such as CORACA that have universal and automatic membership are more likely to include the poorest among their beneficiaries (except for those who are absent most of the year due to migration pressures). Although more equitable, this makes it much more complicated for these organizations to achieve any form of economic sustainability, because they are faced with the challenge of managing service delivery systems in such a way that they serve their poorest members and, at the same time, become self-financing.

Organizing for Change—Organizing for Modernization?

Viability and Sustainability of the Federations

The dynamics influencing the quality of the federations' impact are closely related to the viability and sustainability of the technology support they offer. We have argued that the support is significant only when it is ongoing and linked to the development and negotiation of new market opportunities and relationships. In order for the support to be ongoing, either the federation itself or the mechanism through which the technical support is delivered must be financially sustainable.

Across the cases, there are as yet few examples of financial sustainability. Only FUNORSAL, El Ceibo, UCIG, and CORACA have achieved any real degree of self-financing and then only through certain components of their activities. These are the organizations—or the parts of those organizations—that aim to function as social enterprises through market engagement and product transformation. These activities allow revenue generation both for members and for the organization. Furthermore, the two organizations in this category that are significantly self-financing (El Ceibo and FUNORSAL) have both received support since the early 1980s, primarily from international NGOs, technical volunteers, and the church. This support has included human capital formation, financial and production capital, and significant administrative support. The implication is that long-term, committed, and costly support is needed for a federation to arrive at this level of self-financing and for its natural resource–based activities to begin to catalyze wider local development processes.

The costliness of this investment raises the question as to whether there are alternative means of fostering this form of regional, socially inclusive, intensification process. For the federation to play only a service delivery role is not an alternative since it will not lead to any local multiplier effects. Likewise, there are few cases of financially sustainable NGO-based social enterprises. Therefore, this is not an easy alternative, though it may have a role to play.

Whether the commercial private sector would invest in such a way as to catalyze these processes and ensure that they are socially inclusive also seems unlikely, for two reasons, both related to the fact that these federations function on the basis of "subsidies." On one hand, they need a start-up subsidy (the cost of the donor investment in the first years of the organization). On the other hand, there is a Chayanovian subsidy whereby the staff and members of the organization commit time to the activities of the organization at a rate of remuneration below market rates, because of other nonmonetary "benefits"—that this is their organization, that the work is in their place of origin, and so on. The implication is that, in regions such as the altiplano of Potosí, the high Andes of Salinas, and the high tropical forest of the Alto Beni, these organizational forms are one of the few, perhaps only, potential means of invigorating the economy in a way that is at least reasonably socially inclusive.

Approaches to Pluralistic Participation

Could these federations be players within the conception of "pluralistic participation"—a conception of agricultural and rural development in which a range of actors play their roles and interact with each other in such a way as to foster greater effectiveness and efficiency—as outlined in Uquillas and Pichón (1995)? The cases show clearly that these federations have been willing to interact and coordinate with other institutions in their strategies. The two most common, most successful, and least conflictual types of relationships have been with international NGOs and with technically based public agencies such as research and extension services.

In these relationships, the external organization has provided the federation with technology and technical support. For instance, El Ceibo has accessed planting material from Bolivian and other Latin American research institutes; CORACA has close relationships with the Bolivian national potato program and with the University of Punó in Peru, for animal skin processing; UCIG has links to the national agricultural research institute of Ecuador. Clearly, these are organizations that can be partners with other actors involved in developing agricultural and NRM technologies. Indeed, an assessment of the work of UCIG suggests that the on-farm research program of the national agricultural research institute found it far easier, and far more effective, to work with UCIG than with NGOs.

Nonetheless, it also is clear that such relationships need support. In particular, bridges between the federations and the institutions are necessary to help initiate the relationship. Without these bridges, the links do not happen as easily as, for instance, relationships between service NGOs and national research institutes. This is largely for sociocultural reasons. NGO staff are generally from the same social, cultural, and educational worlds as government and international technical staff and, in some cases, are already acquainted with each other. Conversely, federation leaders are from socially different worlds and have few (if any) points of contact with researchers. In addition, the leaders have little technical legitimacy with these other organizations. Therefore, such organizations are less inclined to take the federation leaders seriously when they suggest collaboration.

In this situation, one common bridge between federation and research institute has been via the employed (usually noncampesino) technical staff of the federations who may often have colleagues and university friends working in these other institutions. In other cases, donor agencies have helped to make the initial contact. CORACA's links to the national potato program in Bolivia, for instance, were greatly assisted by the Dutch NGO Servicio Holandés de Cooperación al Desarrollo (SNV) who contracted an international expert to do a short consultancy at the onset of the project. This consultant had no difficulty in gaining high-level contacts with research institutions, and on the basis of these meetings CORACA now has several collaborative links with the potato program.

Two additional points merit attention. The first is that, in general, the federations have far better relationships with technical (as opposed to political) government institutions than they do with NGOs. Relationships with NGOs are often competitive. Federations argue that NGOs control resources that should be destined for the federations and seek clients among the rural population at the expense of the federations' attempts to strengthen their presence at the grassroots. At other times, there are straightforward political disagreements with the NGOs, who may have particular party orientations that differ from those of the federations.

The tense relationship with NGOs is perhaps more apparent among the federations who have come to technology development from a more political origin. However, as the less politicized federations (such as El Ceibo and FUNORSAL) have gained more force, they too have become far more critical of local NGOs. Indeed, the tendency in all cases is for the federation to begin to demand coordination among the activities of institutions acting in their area of influence—even when the federation is too weak in terms of membership loyalty to achieve this demand. Interestingly, the two cases in which federations have been most successful in demanding coordination have been El Ceibo and FUNORSAL. Notwithstanding their less "political" origins, the local legitimacy that their economic importance has bestowed upon them has enabled them to exercise considerable influence over NGOs and government organizations working in their areas.

This leads to the second point. To what degree can these federations hold local institutions accountable and make them more responsive to local needs? For this to occur, the federation must first be responsive, accountable, and able to identify member concerns. The leaders of many federations are not as accountable in practice as they are in rhetoric, and sometimes agencies who work with them pay insufficient attention to assessing the legitimacy and representativeness of leaders. On the other hand, federations can be structured in such a way that leads to more internal accountability—organizations with more internal working groups with different memberships tend to be more internally accountable, although more complex to manage (Fox 1992). Also, notwithstanding these problems of accountability, federations still tend to be more broadly representative and accountable than NGOs.

In practice, several federations have been able to exercise influence on other agencies because of their economic or political strength. Perhaps the most frequent and effective means of doing this has been to occupy institutional spaces where influence can be exercised, such as interorganizational committees. In general, such spaces cannot easily be occupied by community or user groups. CORACA administrators, for instance, sit on several such committees and working groups in Potosí, and one is vice-president of the departmental seed council. However, in the realm of technology development and natural resource management, this influence has basically been limited to encouraging more coordination among local institutions and improving the federations' and their members' access to the services of these agencies. Federations have rarely pushed institutions to change their research agendas drastically, and aside from asking for more work with native varieties of crops (especially potato), they have not pressed for research agendas that pay closer attention to traditional resource management practices. This, in large measure, reflects the tendency of these federations not to have a particularly critical attitude toward modern technologies.

Conclusions and Questions

To close this reflection on the possible roles of campesino federations in NRM and technology development, let us return to several of the themes introduced by Pichón and Uquillas (this volume): the links between traditional and modern approaches to technology and natural resource management, the possibility of agricultural and natural resource intensification, and the options for pluralistic participation. These themes serve as the basis for three questions that help structure the concluding comments. The first question is, Why do federations of campesinos (most of whom are indigenous people) in practice pay such limited attention to traditional and indigenous technology? The second is, Are federations, in any way, really successful in increasing the possibility that rural people will be able to stay in rural areas by improving their livelihood possibilities—be it through direct intensification of their agricultural systems or through the more general intensification of livelihood possibilities? The third is, What role might these federations play in a "pluralistic" participatory approach to agricultural development (Uquillas and Pichón 1995)?

Tradition and Modernity

In large measure, the internal dynamics of the federations push them toward promoting the technologies of modernization. Although some organizations—primarily those influenced by strong indigenous politics, such as FOIN or UCASAJ—may begin working with traditional technologies, the tendency is for this focus to change over time, largely because pressure from members tends to be for support of new techniques that result in a quick rise in farm-level output or income. Therefore, because organizations often "use" a technology support program as a means of strengthening the organization by establishing credibility at the member level, "rapid impact" technologies are preferred. NRM and traditional technologies rarely achieve the same results, except in special circumstances—for instance, when linked to securing tenure rights (such as through forest management) or to the existence of organic product markets.

Organizations that deliver technology to their members are limited by the technology that is available within the wider system—and this tends to be biased toward modern technology. Thus, CORACA has to work with modern varieties of potato and El Ceibo with exotic varieties of cacao because the wider research system has done little or nothing to improve the quality of native planting materials. As improved native materials become more available, it is likely that the federations will work with these—because they are native and, above all, because they are improved.

Organizations also are market oriented in the technologies they employ because their members are primarily concerned with improving their income through market engagement. Thus, organizations tend not to work with a range of traditional crops that have low market demand. Conversely, they will work with more sustainable and lower external input technological options when there is a special market for these. The best examples of this are El Ceibo's work with organic cacao and the work of other organizations in the altiplano of Bolivia with organic quinoa (organizations such as ANAPQUI and CECOAT). To date, these market niches are small, especially in the domestic market.

Organizations are likely to make significant efforts in working with traditional approaches to resource management and technology development only under certain conditions. One condition is that there are significant economic incentives to their members for engaging in these practices and utilizing these crops. A second condition is that the wider technology development system has already improved the quality of these techniques and crops, particularly in the form of clean genetic material.

Although the technical options promoted by the federations are modernizing, the organizational principles underlying much of this work have long and deep roots in indigenous culture and practice. Thus, running through all this work is a strong commitment to the idea of local control and a strengthened capacity among local organizations and people to negotiate within a wider political and economic system. In Andean tradition, this idea has roots going back before the arrival of Europeans. Similarly, underlying the work of all these organizations is a strong commitment to place, that is, a strong commitment to finding a means for rural people to continue living in rural areas. In the same sense, they have a strong commitment to elements of cultural and social identity that derive from continuing ties to place. This is the deeper logic that underlies much of their work in technical assistance, income generation, and land rights.

Markets and Intensification

This leads to the second question, Are federations really successful in enabling rural people to stay in rural areas, by improving their livelihood possibilities—be it through direct intensification of their agricultural systems or through the more general intensification of livelihood possibilities? Here, we have to be quite frank: the impact has been limited. In those cases where federations have worked only in service delivery, the contributions have been marginal. The volume of support delivered to any one family is minimal, and aside from the genetic improvement and diversification of planting material that might derive from a seed program, the impact is very short-lived.

We also must acknowledge, however, that in this respect the federations have fared no differently from NGO or government interventions, which have had limited impact on rural livelihoods (see van Niekerk 1994 for recent cost-benefit assessments of NGO programs in the Peruvian and Bolivian altiplano). Indeed, we might well argue that those federations following the path of social enterprise have much more to contribute to agricultural intensification than many NGO or government programs. Among the cases reviewed, it is only in El Ceibo and FUNORSAL (and marginally in CORACA) that there has been any significant contribution to rural livelihoods, and these are the only cases where the contribution might be sustainable. These organizations began with a marketing and processing role and have subsequently moved into technology development and natural resource management, in order to increase or protect the capacity of their members to exploit new market niches.

By opening new markets and by transforming and marketing produce, these cases point to a means whereby economic incentives for the sustainable intensification of natural resource management can be created and rural livelihoods can be enhanced more sustainably. Technology per se will never be enough. However, these cases also remind us that building such organizations requires time, resources, and commitment.

Potential Roles of Campesino Federations

The third question is, What role might these federations play in a pluralistic participatory approach to agricultural development (Uquillas and Pichón 1995)? At the outset of this chapter, we stressed the risk of romanticizing these organizations as traditional knowledge is often romanticized. Thus, when thinking of the potential role of these federations in any interinstitutional rural development strategy based on natural resource intensification, we must be cognizant of the weaknesses of these organizations. Indeed, if the federations were asked to perform many additional roles too quickly and without adequate preparation, this could do more harm than good to the organizations.

In thinking about potential roles of such organizations, we should begin not with the organizations but with current processes of change in the region. There are some critical trends in this regard: the increasingly serious livelihood crisis facing many, maybe most, Andean communities; the continuing failure of private capital to invest in any significant employment generation in Andean small town and rural economies; and the institutional transformations occurring as a result of the reformed role of the public sector in the region. It seems evident that the only way livelihood options are going to improve is through economic organization and activity that enhance rural wealth creation and that increase how much this wealth is reinvested in income, employment, and further wealth-generating activities in local rural economies. As long as value is not being added to, or is at least latent within, agricultural production and natural resources in situ or on an extractive basis, local economic development will not occur and the creeping livelihood crisis will continue.

Looking at the range of actors in the Andes, it would appear that those with the greatest proclivity to reinvest within the region are campesino economic organizations. Most forms of private capital seem unlikely to fill the void (with the exception of some regional capital), and those that are willing to invest private capital will still extract a certain share of value added for consumption and capital costs. The NGO sector has a very poor record in the area of productive activity and social enterprise. Given the urban origin of NGOs, they also would extract a proportion of any value added in the form of wages and administrative costs, as is already the case in their use of project grant resources. State direct investment is clearly a thing of the past in Latin America—at least for the foreseeable future.

This would appear to leave individual and organized farmers as the only actors likely to engage in such investment, but here too the record is not that positive. Examples such as El Ceibo and FUNORSAL are as significant for their rarity as they are for their success. As several of the cases show, however, there are seeds of possibility in other organizations. In the current climate, one role for these federations will be in the area of natural resource–based social enterprise, and this points to an important area in which links with other organizations could be forged. Multiple forms of support are needed, and these define important areas where NGOs and donors can contribute. There are particularly critical areas: market identification and development; developing capacity in social business administration; production transformation, marketing, and other forms of value aggregation; and appropriate forms of human capital formation. Areas where technology development and NRM improvement might lead to significant benefits would then derive from the market options identified by these organizations. The examples show that federations are both willing and able to develop the necessary links with research and other institutions to respond to these technology challenges.

Another clear, related role for the federations is to link rural areas with a wider technological and institutional environment that cannot easily be managed by traditional family- and community-level institutions. Once again, federations will need some support in this endeavor, above all to initiate the contacts with external institutions. Initially, this role need not involve the federations' engaging in enterprise activities —but, this chapter argues, in due course this will be vitally important for the creation of new and improved livelihood options.

Finally, the traditional pressure group role of federations will continue to be important. In a policy environment that is increasingly dominated by a handful of neoclassical concepts, and in a political environment where the legal controls on the excesses of private capital are being weakened, it will remain critical that there are institutions to give a voice to the rural poor and to help defend their natural resource entitlements and rights. For a variety of reasons, NGOs and political parties are imperfect in this regard, and though far from perfectly representative themselves, the federations do offer the possibility of representativeness. This will be another area where the federations could benefit from support. So in this new environment of pluralistic participation, federations will continue to have a slightly more political and confrontational role to play—as well as their pluralistic partnership roles. But then, participation always was as much about negotiating different and conflicting interests as it was about being friends.

5

Organizing Experimenting Farmers for Participation in Agricultural Research and Technology Development

Jacqueline A. Ashby, Teresa Gracia,
María del Pilar Guerrero, Carlos Arturo Quirós,
José Ignacio Roa, and Jorge Alonso Beltrán

Farmers who experiment are an important resource in helping rural communities solve their farming problems. Yet, these experimenting farmers are generally unrecognized and unsupported, disconnected from often substantial investment in formal agricultural research. Experimenting farmers as a resource are neglected because conventional approaches to agricultural technology generation are top-down. Technology is designed by scientists who make decisions about what to recommend to farmers without giving the farmers any direct say in the process. The conventional approach is like a doctor-patient relationship. The researcher and extensionist (doctors) are supposed to formulate a prescription to cure the ailments of the farmer (patient). But when the doctor or scientist neither correctly diagnoses a sufficient number of problems nor formulates appropriate prescriptions because needs are so many and diverse, this approach breaks down. Developing technology that is suited to the particular location-specific needs and problems of the 1.5 billion people who depend on complex, diverse, and risk-prone agriculture requires a different tactic (Chambers 1994).

One solution might be the establishment of a community-based capacity to carry out adaptive research, with the participation of farmers in identifying problems and in implementing technology testing. There is increasing information on a growing number of instances of organizing groups of farmers or working with existing farmer organizations to implement farmer participation (see, for example Drinkwater 1994a; Heinrich, Worman, and Koketso 1991; Mattee and Lasalle 1994; Muchagata, de Reynal, and Verga 1994; Mushita 1993). The strategy of organizing groups of farmers to participate in adaptive technology testing is, in part, a response to concerns about how to reduce the costs of including farmers in research when this makes heavy demands on the time of salaried professionals such as researchers or extensionists. This strategy also addresses the need to "scale up" farmer participation in research and extension, so that technology testing can be carried out in numerous, diverse microenvironments without incurring excessive expenses and compromising the quality of participation (Ashby 1990; Bebbington, Merrill-Sands, and Farrington 1994; Okali, Sumberg, and Farrington 1994).

Many questions have been raised about the viability of institutionalizing an adaptive research role for farmers as well as the constraints such efforts are likely to face (Bebbington, Merrill-Sands, and Farrington 1994). Critics argue that farmers' traditional or folk experimentation is a form of knowledge generation that is superior to Western science. The strength of folk experimentation is that it makes contingent, sequential adjustments over time to circumstances that change unpredictably (Drinkwater 1994b; Richards 1989; Scoones and Thompson 1994). This indigenous form of knowledge generation does not readily fit within the models of controlled experimentation employed by Western science.

A more useful analysis draws on understanding the nature of folk experimentation. For example, farmers compare "treatments," but the check or control may be "in the farmer's head," because farmers compare this year's performance with last year's. Another comparison used by farmers contrasts results in a distant field with one nearby; or the results of adding a little bit more fertilizer to one furrow compared to the remainder of the field. Folk experimentation involves replication, but typically replication over time, in contrast to the replication in space and time that is characteristic of the scientific method. Moreover, farmers recognize confounding effects in folk experimentation. For example, a small amount of seed of a new variety is typically nurtured and propagated in the more fertile home garden; the next planting then moves the new variety to testing in different types of soil, thus testing the genotype x-environment interaction. Only after assessing performance in a variety of microenvironments are conclusions drawn about the likely performance of new germplasm in the farmers' environment. Experience shows that farmers' knowledge generation can draw both on the scientific method of controlled comparison and on folk experimentation: it is not an either/or dichotomy (see Berg 1993; Hardon and de Boef 1993:67; Lightfoot 1987; Uphoff 1992:282–83).

Another issue is whether the creation of a special group builds on existing authority structures or creates a parallel nontraditional structure. Also at issue is the extent to which such groups can represent the research agendas of different interest groups within the community or may indeed exclude particular groups (Bebbington, Merrill-Sands, and Farrington 1994). Experience with on-farm research shows that, when formal criteria for selecting farmers to participate in research were not used, participants usually ended up being the more wealthy and politically active farmers (Merrill-Sands et al. 1991:303).

Experimenters (or innovators who can afford to experiment) are likely to be the relatively better-off farmers who have the skills and resources (including power) to devote to a particular kind of knowledge generation. Some instances indicate that, when working with research-minded farmers, it is desirable to purposely select innovators who have the time and interest for experimentation (see Abedin and Chowdry 1989, cited in Merrill-Sands et al. 1991; Ashby, Quirós, and Rivera 1987). The real issue is whether experimenting farmers who represent the local capacity for research in rural communities can be harnessed to a research agenda that is defined at the community level and that also is useful to the very poor or to other interest groups (such as women) who may have priorities that vary from those of the relatively better-off local experimenters.

Could this local capacity, if linked effectively to research agencies, share the costs and expand the coverage of adaptive research while ensuring that this is relevant to local farmers? How is community-based participatory research to achieve broad coverage that is cost-effective? Can a self-sustaining capacity and responsibility for promoting farmer participation be created in rural communities? How can linkages among these different actors be managed without increasing the transaction costs to a ponderous level? Little systematic work has been done on the costs of creating organizations at the community level to fill this function. Nor has there been much empirical assessment of the extent to which such organizations can increase coverage and improve targeting of adaptive research in a way that is self-sustaining (Axinn 1994).

This chapter reports on an effort to provide empirical data on some of these issues from action research carried out between 1990 and 1994 by the Investigación Participativa para la Agricultura (IPRA) project of the International Center for Tropical Agriculture (CIAT), with support from the W. K. Kellogg Foundation. The IPRA project aims to assess the potential for institutionalizing a community-based capacity for involving farmers in carrying out adaptive research. In this chapter we report results obtained on the devolution of adaptive research responsibilities to committees of experimenting farmers, the effects of upscaling this approach to achieve broad coverage, and the related costs. The chapter is organized into three main sections: description of the procedures used for forming farmers' committees and their activities; results in relation to the evolution of the farmers' committees over the four-year period from 1990 to 1994; and issues for the future application of this approach.

Methodology

The project's strategy is to implement participatory research methods for adaptive technology testing by forming committees of farmers based in rural communities. The committees will carry out technology testing with public sector agricultural research and extension agencies and intermediate organizations such as NGOs and farmer cooperatives. Development of courses and materials used for training farmers and the staff of public sector and intermediate organizations for this purpose is integral to the strategy.

Organizing Experimenting Farmers

The purpose of the farmers' research committees, called Comités de Investigación Agropecuaria Local (CIALs), is to mobilize local leadership among farmers, and for the leaders to take responsibility for experimenting with technologies not yet known in their community. In this way, the project aims to create "demand-pull" by clients of public sector and intermediate organizations for agricultural research and extension; diversification of the types of technologies available; and increased number and rate of flow of technologies to resource-poor farmers in order to improve adoption, farm incomes, and welfare. Experience shows that when new technology is selected with farmer participation methods, it is better adapted to local conditions than when recommended by researchers working on their own (Worede and Mekbib 1993).

The project was initiated in a pilot area in Cauca Department, southern Colombia. Cauca is one of the poorest, lowest-wage departments in the country. The pilot area is characterized by hilly terrain, poor infrastructure of roads and markets, and small farms averaging five hectares in size (average cultivated area is less than three hectares). All farms engage in a mix of commercial and subsistence production. It is a marginal coffee-production area with infertile acid soils, often badly eroded. Most farmers cultivate coffee and cassava as cash crops, with some maize and climbing beans for subsistence. Livestock is scarce (only 13 percent of farms have any cattle).

The project began the formation of CIALs in five communities (veredas) in 1990. This number increased to eighteen communities in late 1991, thirty-two in 1992–1993, and fifty-five communities by 1994. An additional thirty CIALs, formed in Bolivia, Ecuador, Peru, and Honduras by international trainees in the method, brought the total to eighty-five in 1995. The information presented here was obtained from monitoring the forty-eight CIALs formed between 1990 and mid-1994 in the pilot area in Colombia. These CIALs cover an area of approximately 1,605 square kilometers and incorporate an estimated 50,000 families, with over 4,000 farmers directly involved in the project, 220 of whom participated in training as members of the CIAL or research committees.

Each CIAL is formed by four farmers, who are elected at a community meeting and who meet regularly during the first training cycle or experimental period (usually equivalent to a cropping season of about six months). The first training cycle involves up to ten training visits by a support-farmer who has had at least one year of prior experience in a CIAL. Over the next cycle or cropping season, these visits are progressively reduced in number as the new CIAL gains experience and carries out experiments with increasing autonomy (see appendix 1 at the end of this chapter). The support-farmer is, in turn, assisted by an agronomist who provides input to the statistical design of CIAL experiments and the analysis of data collected by CIAL members. At present in the project area, the forty-eight CIALs are attended by three support-farmers, served principally by one trainer-agronomist.

Results

It is necessary now to review the results obtained from 1990 to 1994 from the organization of forty-eight CIALs or farmers' research committees in the pilot area in Cauca, Colombia. The procedure for forming CIALs was developed in a pilot phase from 1990 to 1991. Five farmers' research committees were established and their members trained in techniques for participatory diagnosis, planning and establishing replicated on-farm trials, participatory evaluation of technology, analysis and interpretation of results, and budget analysis of the total cost of the trial and of the individual treatments. Planning and presenting a short oral report on the results to each community and to a meeting of all five CIALs was part of the process. After the first training cycle or cropping season, the agronomists in the IPRA project team began gradually to turn each operation in the process over to the farmers. A sociologist made regular monitoring visits to assess how well the farmers could manage each operation and to detect when follow-up training was required. On the basis of this experience, training materials—in the form of twelve CIAL handbooks—were prepared in discussions with the farmers involved, who helped prepare the text and illustrations (see the IPRA materials cited in the bibliography of this volume).

At the end of 1991, the second phase of CIAL formation was initiated. The project used the training materials to teach a course with NGOs in the pilot area in order to prepare their staff agronomists to establish CIALs. An additional thirteen CIALs were established using the training handbooks. Monitoring by the project of this second phase now covered eighteen CIALs and included revision of the training handbooks, as these were used in practice by the NGO trainees and their CIALs. Based on this experience, the training handbooks were finalized, and in 1992 the project began on a regular basis to teach a course on the CIAL method to NGO trainees (university students doing a six-month agricultural extension practicum in the rural areas with the NGO), state extension agents, and local community leaders.

In the third phase, another twenty-eight CIALs were formed, in response to requests from communities and farmers' associations. In 1993 trainees in the course on the CIAL method included three farmers who were members of CIALs formed in the second phase. These support-farmers were contracted (one by an NGO, one by a farmer cooperative, and one by the IPRA project) to form the CIALs in the third phase.

Devolution of Responsibility for Adaptive Testing

One of the most important questions for the project is, What types of responsibility for location-specific technology testing can be successfully assumed by experimenting farmers organized in a CIAL? The forty-eight CIALs formed in the pilot area were established successively by trainees new to the methodology. This has enabled the project to evaluate the training requirements for establishing new CIALs, the rate at which CIALs can be progressively "detached" from their trainer, and the rate at which they can take over responsibility for carrying out experiments in the absence of a trainer.

Table 5.1 summarizes conclusions on the type of institutional support required by a developed CIAL in the form of training and monitoring visits to carry out a crop-related, on-farm experiment. Our experience demonstrates that, when they are interested in the results, farmers' committees working on their own accurately record results for separate treatments. When farmers have defined data they want to collect (in measurement units that make sense to them) in planning meetings, they are able to analyze these data to compare treatments and to assess germination rates and crop development. For example, in evaluating

Table 5.1. Activities Carried Out by a CIAL in a Crop-Related Experiment and Institutional Support Required

Type of Activity	Number of Activities	Training and Assistance by Agronomist	Training and Assistance by Paraprofessional	Monitoring by Paraprofessional Farmer
Group diagnosis	1	—	—	(1)*
Planning	1	1	—	—
Plot selection	1	—	1	—
Land preparation	1	—	—	—
Obtain inputs	1	—	—	—
Establish experiment	2	—	—	—
Check germination	2	—	—	—
Crop management	variable	—	—	—
Midterm evaluation	2	—	1	—
Harvest evaluation	1	—	—	—
Analysis of results	1	1	—	—
Community report	2	—	—	1
Total	15	2	2	2

Note: *Infrequent

the height of maize plants, the following measures were used: too low —dogs can reach and steal cobs; medium (desired)—will withstand wind and resist lodging; too tall—susceptible to lodging. Yield data are commonly processed by farmers in terms of yield per unit of seed because they do not customarily use measures of area, even though experimental plots are measured and staked out.

Monitoring of the forty-eight CIALs shows that, of the fifteen activities listed in table 5.1, a fully developed CIAL with two cycles of experience required training support in, at most, four activities. In phases 1 and 2, two of the activities (planning, including the statistical design, and analysis of results) required the presence of a trained agronomist. For the other two activities (checking that plot selection is consistent with the experimental objectives and ensuring that data at the midterm or harvest evaluation are collected), visits by a support-farmer were identified as desirable. Monitoring visits routinely involve a visit by a support-farmer to the community meetings for diagnosis and reporting. By phase 3, support-farmers begin to take over responsibility for assisting in planning the CIALs' trials, analyzing results, and presenting plans (and later, results) to a meeting with one of the host institutions' agronomists. Most of the CIALs' research questions can be addressed by single factor experiments—for example, from six to ten varieties superimposed on local cultural practices, or 3-4 fertilizer application, or pest control treatments. This makes support of trial planning and analysis feasible for paraprofessional farmers in conjunction with an agronomist who verifies the design and interpretation of results.[1]

The research agenda defined by the CIALs is evolving from primarily germplasm-based strategies (the search for new crops and varieties) to an interest in cultural practices after a viable new crop or locally adapted varieties have been selected through their experiments. Recent experience shows that the paraprofessional farmers can support CIALs in the design of a two-factor experiment, such as planting density x fertilizer dosage, without the intervention of an agronomist.

Table 5.2 shows the rate of increase in activities carried out by CIALs, independent of institutional support in the form of training by the agronomist or support-farmer. The data on the first cycle show that the number of training visits required has been reduced from seventeen (needed to develop the method and the training materials) in phase 1 to ten in phase 3. Training sessions follow the activities outlined in table 5.1. In practice, the number of training visits has been reduced, because some activities listed in table 5.1 (such as obtaining inputs and regularly scheduled observations or evaluations of the experiment) can be carried out independently, by the farmers, even during the first training cycle.

Table 5.2 shows that the newest CIALs (formed in phase 3) are, by cycle 2, operating with an average of four visits for training and support. The five pilot CIALs (formed in phase 1) have continued to increase their autonomy. By the last cycle, these pilot CIALs were operating with only two support visits (one for planning the experiment and one for the analysis of results) by the paraprofessional farmer.

To conclude, the project's experience demonstrates that the training of farmers' research committees can be accomplished in two cycles (that

Table 5.2. Decreasing Dependence of CIALs on Institutional Support, 1991–1994

Phase of CIAL Formation	*No. of CIALs*	*Mean Number of Visits per Training Cycle*				
		Cycle 1	*Cycle 2*	*Cycle 3*	*Cycle 4*	*Cycle 5*
Phase 1 (pilot)	5	17	10	4	2	2
Phase 2	13	12	7	3	3	—
Phase 3	30	10	4	4	3	—

is, during two experiments) and that such fully trained CIALs can take over responsibility for the majority of the activities required for farmers to implement on-farm experiments. Phase 3 experience (1993–1994) indicates that support-farmers (with two cycles of experience as a CIAL member) can provide almost all of the training and monitoring required for formation and maintenance of the CIALs' experimentation in the form of simple on-farm trials. This allows the agronomists involved in the project to delegate the planning and analysis needed to routinely support fully trained CIALs, as well as formative activities for developing CIALs, to support-farmers.[2]

Quality of Research Conducted by CIALs

The next issue of importance is the quality of the research carried out by those CIALs that operate with the degree of autonomy discussed above. Evaluating the quality of farmers' research is related to the issues discussed earlier—especially the usefulness to the farmers of the scientific method as compared to folk experimentation. The project's strategy is to combine both approaches. That is, a formal experiment is planned and established, but if farmers decide to make changes in treatments or to alter the experiment along the way in the style of folk experimentation, then the only requirement is that this decision is made in committee by the collaborating farmers.

CIAL experiments have been established with a minimum of three replications and, on occasion, also are replicated within each site if an agronomist judges this to be advisable. Trials have taken place on land belonging to members of the CIAL, on communal land, or on land belonging to other farmers, and have included rental or sharecropping arrangements common in the community in question. Site selection is a decision made by the CIAL, with a follow-up visit to check that the proposed sites are consistent with the experimental objectives identified in the planning activity. Consequently, the trials are carried out under a variety of collaborative arrangements involving, on occasion, a group of community members who donate labor for the trial or an individual who sharecrops and contributes land or labor for a share of the harvest. Observation and evaluation of the progress of the trials may involve several experimenting farmers—identified by the CIAL committee members as knowledgeable experts in the topic chosen in the community's diagnostic meeting—who then take part in planning and implementing the replications. There exists, therefore, space for farmers to intervene and to combine folk experimentation with the formal experimental design.

The project has monitored the quality of CIALs' research with respect to three criteria, best presented in the form of questions. (1) Is the experiment interpretable by farmers and also statistically analyzable? (2) Were farmers still satisfied that they could draw useful conclusions from the experiment, even if statistically unanalyzable? (3) Did farmers conclude that they could not draw useful information from the experiment? The evaluation consequently asks if farmers perceive the experiments to be useful for generating information, as well as whether the experiments have the potential to supply useful information to formal research and extension systems.

Table 5.3 presents the results of this evaluation. Of the 273 trial plots managed by CIALs between 1991 and 1994, the percentage of plots (replicates) that could be used for statistical analysis averaged 75 percent. In phase 1, 91 percent of plots were judged interpretable by farmers, although somewhat fewer (84 percent) were statistically analyzable. In phase 2, farmers still judged 89 percent to be interpretable for their purposes, although only 62 percent were statistically analyzable. The reasons for this drop in percentage of statistically analyzable data were found in the self-evaluation exercise conducted with each CIAL. In the second phase, CIALs were linked to trainee extension agents who were managing the supply of inputs for the CIAL experiments together with those for their NGOs credit program. The credit program was plagued by delays in obtaining funds for purchasing in-

Table 5.3. Quality of On-Farm Trials Conducted by Farmer Committees (CIAL), January 1991–August 1994

Phase of CIAL Formation	No. of Plots	Percent of Plots Statistically Analyzable and Interpretable by Farmers	Percent Statistically Unanalyzable but Interpretable by Farmers	Percent Lost to Analysis
Phase 1 (pilot)	42	84	7	9
Phase 2	85	62	27	11
Phase 3	146	78	12	10
Total	273	75	15	10

puts given in-kind to participating farmers. This resulted in planting delays and subsequent loss of trial plots from the CIAL experiments. The CIALs requested that they manage the petty cash fund for purchasing experimental inputs. Once this was put into operation, the capacity of the CIALs to implement their trials in a timely fashion improved significantly. In phase 3, the number of plots lost to analysis because of late planting decreased to three (managed by one CIAL), with other factors accounting for the lost-to-analysis of the seven remaining plots. In sum, the average success rate by statistical criteria is 75 percent. The average success rate by farmers' criteria—in terms of carrying out trials judged locally useful for knowledge generation—is 90 percent.

Reasons That Devolution Succeeds or Fails

Why is this degree of responsibility and accuracy in conducting adaptive trials achieved by resource-poor farmers who are very busy people, struggling to cope with running their own plots and farms? This "success" is especially puzzling in view of the huge resources devoted in the past to training and equipping teams of on-farm (or farming system) researchers for whom obtaining farmer collaboration or participation in formal experiments was a major source of frustration (Lightfoot and Barker 1988). Analysis of the success and failure of CIALs over the period from 1991 to 1994, during which time five CIALs have become inactive (representing 11 percent of the total number formed), suggests that there are several determinants of the degree to which a CIAL makes a commitment to running its experiments with a minimum of institutional support.

First, the CIAL's training must successfully impart the principle that the committee's objective is to experiment in order to generate knowledge and disprove or discredit unreliable recommendations. If this objective is not clear, CIAL members experience a loss of purpose if an experiment shows that local practice is, in fact, the best available alternative for the innovation being tested. Our experience shows truly impressive persistence on the part of some CIALs in the face of several experiments that did not identify a promising innovation when compared to local practice. In this respect, contact among CIALs is an important ingredient of success—one CIAL benefits from another's experimentation and is motivated by it (see Appendix 2, p. 182).

Second, it is obviously useful for a CIAL to include one literate member who can read the CIAL handbooks aloud to the other members and who can keep records and tally accounts, because this facilitates the management process. However, written records have proved important mainly to the host institution that is collating data from several CIALs. Nonliterate farmers recall complicated varietal code names, the schedule of treatments, and the differences between treatments with amazing facility if they perceive the information as important and useful to them. Our experience suggests that literacy may not be a prerequisite for farmers to carry out the CIAL's adaptive research responsibilities, but it does mean that support by the paraprofessional may have to be more intensive over an extended period. Nor does functional illiteracy prevent farmers from exchanging results with each other, since an oral tradition is strong in these communities.

More critical to success is identification by the group of a problem or question, within the framework of a CIAL experiment, that the farmers want to answer and that is of interest to the community. This is why the monitoring visits at diagnosis and report times are important. They ensure that the committee feels accountable to its community and, at the same time, gets encouragement from the interest shown in its results. The sense of community service and responsibility to group welfare created and reinforced in these meetings is possibly the single most important determinant of successful completion of the experiment by a CIAL. Monitoring shows that conflict in the community or conflict in the CIAL is most likely to result in an inactive CIAL. For this reason, the project's approach includes periodic evaluation by the CIAL in terms of how its members feel about each other and their relationship to their community.

Another motivating factor is that managing a CIAL and its experiments commands the respect of outsiders and this has proved a useful tool for attracting the attention and resources of institutions external to the community. For example, one CIAL has successfully negotiated a grant of land for a communal farm from the state land reform agency on the strength of demonstrated management capacity and teamwork. Others, because of the quality and quantity of produce resulting from experimentation, have attracted marketing arrangements with middlemen who previously would not journey to a distant village. Still others have persuaded the NGOs to introduce results of CIAL experiments into their credit programs. The motivation to run an experiment autonomously is related as much to its organizational function as an interface with external organizations as it is to its usefulness as a method of knowledge generation.

Impact of CIAL Trials

The results reported so far show that the forty-eight CIALs in the pilot area have, with a decreasing amount of institutional support, carried out a large and increasing number of on-farm trials that farmers consider useful for knowledge generation and that are to a large extent statistically analyzable. A rapid appraisal of the CIALs' impact shows that in 75 percent of the participating communities there was a perceived benefit from the CIAL in the form of new seed, new cultural practices, or information on recommendations that should be followed. Of those CIALs with no perceived benefit, all but two were formed in phase 3 and are, therefore, newcomers less likely to have experienced positive impacts so far. One community, for example, asked its research committee to compare the state agency's recommendation (to cover the ground under fruit bushes with black plastic) to a local practice for controlling nematodes. The CIAL experiment shows that, at least to date, the local practice is more effective under farmer management.

In the pilot area, state institutions set research and extension priorities on the basis of area devoted to different crops in the municipality. Thus, in the pilot area, the priorities are cassava, pastures, sugarcane, and coffee. Small farmers participating in the CIALs' diagnostic meetings had different priorities, as is illustrated by the crops selected for CIAL experiments listed in table 5.4. It is apparent that the communities identified a more diverse research agenda than the institutions. Not one community prioritized cassava in their group diagnostic meetings, for example, even though over four thousand farmers have participated in community meetings for this purpose. Diversity in the CIAL agenda reflects farmers' objectives to identify alternatives to traditional cash crops (coffee and cassava) and to increase their food sufficiency by growing staples such as potatoes, beans (a substitute for meat in the rural diet), and maize, which is used for feeding chickens, an important source of locally produced protein as well as an important ingredient in traditional staple dishes. In the Cauca Department, these staples are imported from other parts of the country to meet food requirements, so the local food self-sufficiency agenda reflects a regional problem.

At present, the CIALs are having some success in exerting demand-pull and improving the diversity of technologies offered via state and NGO programs. In the NGO credit and technical assistance programs, for example, maize and peas were introduced as a result of CIALs' experiments, and beans were given more importance. The municipal credit and technical assistance agencies (called UMATAs) also began to respond to farmers' priorities, especially once some of the CIALs started producing seed of varieties they had selected in their trials. UMATAs recommended and distributed these seed varieties to other farmers. One UMATA recently began to use the CIALs' results for formulating recommendations to farmers participating in its credit program. Another responded to a CIAL's request for help in identifying peanut varieties for testing by obtaining a selection of new varieties from ICRISAT through the national agricultural research agency.

Another important development was the evolution of some of the CIALs, which had successfully selected new locally adapted crop varieties, into small seed-production enterprises that delivered the new seed and seed of local varieties to farmers in the area. To date, six CIALs have

Table 5.4. Experiments of Local Agricultural Research Committees

| Topic | *Number of Experiments*[1] | | |
	Phase 1	Phase 2	Phase 3
Peas and related cultural practices	1	3	5
Potato	6	5	3
Maize and related practices	7	8	7
Peanut	1	3	4
Fruits and related fertilizer			
dosages, pest control	3	6	11
Beans	6	4	0
Snapbeans	0	1	1
Tomato	1	1	1
Soybean	1	2	1
Sugarcane	1	1	1
Vegetables	4	7	6
Chicken feed mixes	1	1	0
Forage grasses	1	1	1
Cover crops (green manure)	0	0	1
Guinea pigs	0	0	1
Total	32	42	43

Note: [1]Sums to less than the number of CIALs because not all CIALs are establishing new experiments.

begun to produce seed from six varietal trials (with twenty-three replications) conducted over three years. These CIALs have progressed to commercial-scale plots for this endeavor and have received additional training in simple seed production, processing, and quality-control techniques from the agronomist-trainer. This seed can be sold with state approval as "farmer-improved seed" through endorsement by the national agency responsible for seed certification.

Table 5.5 shows the amount of seed produced by the six CIAL seed enterprises. The CIAL seed is distributed locally in village stores and at weekend markets. An estimated 281 hectares of maize, 3,064 hectares of beans, and 3.5 hectares of field peas (an entirely new crop introduced by CIAL experimentation into the pilot area) have been planted with CIAL seed. More than ten thousand farmers have purchased CIAL seed, which, over one planting season, is estimated to have produced yields with a gross value of over two million dollars.[3]

Table 5.5. Seed Production by Six CIALs and Its Estimated Impact over One Planting Season

Crop	No. of CIALs	Total Seed Production (kg)	Estimated Area Planted (hectares)	Estimated Production (tons)
Beans	2	147,080	3,064	3,064
Maize	2	8,430	281	1,124
Peas	2	136	7	3.5
Total	6	155,646	3,352	4,191.5

Crop	Farm Gate Price ($/ton)	Gross Value ($000)	Increment in Production (%)	Value of Increment ($000)
Beans	683	2,093	30	628
Maize	488	549	25	137
Peas	2,439	8	100	8
Total	—	2,650	—	773

Based on the yield differential between locally available varieties and those selected by the CIALs for seed production, this production represents an additional $765,000.00 of gross income to local farmers from maize and beans, and a completely new income source worth over $8,000.00 to date from peas. On a per capita basis, this represents an increment in income worth about one month of wages during one planting season to the farmers who purchased CIAL seed.[4] The seed enterprises also generate employment since they must hire additional labor to plant, harvest, sort, clean, and pack the seed in one-to-five-kilogram sacks (made locally by women). The bean seed enterprises have, for example, generated an average of twenty thousand days of employment locally over five seasons. This is worth an estimated eighty-five thousand dollars at current wage rates over the five years of operation.

This impact has been achieved by six CIALs formed early in the project. Of course, there is no guarantee that the newer CIALs will replicate this experience by identifying new practices or germplasm with comparable impact. The six CIALs that have developed into seed enterprises may have already captured the best opportunity and the windfall profits from participatory breeding and seed production. The impact of the newer CIALs may be more difficult to realize, especially as the emphasis of their research agenda is shifting from grains to perishables (see table 5.4). On the other hand, the recent introduction of field peas via CIAL experimentation suggests that the potential for significant impact from CIAL experimentation with high-value crops exists.

Scaling Up and the Costs of the CIAL Program

The results presented to this point show that a fully trained CIAL can take responsibility for executing most of the activities involved in the management of the adaptive research trials required for the research agenda identified by the forty-eight participating communities. The experimental results have been useful for knowledge generation and have helped to increase the diversity of technology tested and to improve the rate of flow of technologies to the participating communities. This has resulted in sizable monetary benefits in the case of CIAL seed producers.

One of the most important questions addressed by this research was to what extent this type of farmer participation in research could be scaled up to achieve broader coverage, and at what cost. Before participatory research became fashionable, critics often queried whether this approach was an expensive luxury—attractive on a case-by-case basis when supported by highly skilled professionals but not affordable for working with large numbers of farmers (Farrington and Martin 1988). We shall now present information on the potential of the CIAL method to increase the efficiency of salaried personnel working in on-farm adaptive research and extension by decreasing the amount of time required to carry out on-farm trials. We also examine the operating costs of the CIAL Corporation (a second-order organization formed by the CIALs in Cauca) to provide insight into the feasibility and costs of creating self-sustaining CIALs.

One way of assessing the potential of the CIAL method to increase the efficiency of public sector or NGO programs carrying out adaptive research is to compare the amount of time required to conduct an on-farm trial with and without a CIAL. Table 5.6 presents estimates of the workdays required and the cost of labor for an on-farm trial run by an

Table 5.6. Comparison of Labor Requirements of an On-Farm Trial
Managed by CIAL and by Extension Research

Trial Management	Workdays Required[1] (N)	Total Cost of Salaried Labor[2]
By extension research	8	62
New CIAL (Cycle 1)	11	46
Fully trained CIAL	5	23

Notes:

[1]Excludes crop management after trial establishment, which is variable depending on the crop and initial diagnosis.

[2]Support-farmer time costed at minimum wage; extension agent costed at twice the minimum wage; agronomist time costed at average salary current in the pilot area.

extension agent, a new CIAL in the first cycle of training, and a fully trained CIAL. The analysis is based on the activities in table 5.1 and an estimate that an extension agent would require eight workdays for a trial with up to three replicates (sites). A new CIAL requires an average of ten training visits by the support-farmer plus one day of extension agent input to do the same job. A fully trained CIAL can carry out a trial with four days of training from the support-farmer and a fraction of the input from the extensionist (conservatively costed here at one day). Estimates of the different labor costs show that even training a new CIAL to carry out an on-farm trial is less costly than running a trial with a salaried professional—given the pay differentials for the pilot area. More important, devolving an on-farm trial to a fully trained CIAL costs 60 percent less in labor costs than running a trial using an extension agent.

One of the implications of this analysis is that, by working with CIALs, adaptive research programs could potentially reduce their labor costs significantly (by up to 60 percent) for on-farm testing. On the other hand, a given amount of professional labor can be expected to at least double the coverage, that is, to increase the number of on-farm trials and farmer groups served by working with CIALs. Important variables that affect the efficiency and coverage of adaptive research are the variability of microagroecological regimes, the density of the population, and the type of terrain (which affects the amount of time required for site visits). Before examining the current operating costs of the CIAL Corporation, we need to describe the sociogeographical context wherein the existing CIALs have been developed.

In early 1995, there were fifty-five CIALs in existence. These were scattered over nine municipios in the Department of Cauca in southern Colombia, covering an area of 6,648 square kilometers with an average population density of 40 people per square kilometer. The communities of small farmers participating in the project represent an area of influence of approximately 1,605 square kilometers, with a population concentrated at a much higher density than the average. Farm-level surveys show an average of 132 people per square kilometer when extensive cattle and forest holdings are not included.[5] Communities are characterized by land use of 0.25 hectares of cropland per capita—a figure comparable to estimates (see Pachico, Ashby, and Sanint 1994) for Bolivia (0.33 hectares), Ecuador (0.25 hectares), and Peru (0.17 hectares).

Since 1991, the CIALs in the project area have decided to meet on an annual basis to share results. In this one- or two-day meeting, financed by raising money in their communities for transportation and lodging, CIAL members give oral reports on their experiments, exchange seed, swap notes about their host institutions, and formulate recommendations on how to improve their performance on goals they themselves establish in each meeting. This experience prompted the election of a central coordinating committee or junta in 1993 and then, in 1994, led to the CIALs' decision to incorporate legally at the recommendation of the junta. Donations were obtained that enabled the CIAL Corporation to establish an investment fund from which the corporation can draw up to 70 percent of the interest for operating expenses. The remainder goes back into capital. This put the CIALs on a self-sustaining financial basis.

In addition, the support-farmers have begun to give courses to the UMATAs (municipal extension services), which have contracted the support-farmers to form small numbers of pilot CIALs elsewhere. The UMATAS pay up to 50 percent of the support-farmers' salaries for this work. Additional income is thus generated for the corporation. The following data on costs are taken from the CIAL Corporation's annual operating budget, which in turn is derived from the project's data on the costs of running the CIALs in 1994.

Table 5.7. Annual Operating Costs of the CIAL Corporation for Fifty-five
CIALs

	Annual Costs (US $)
Personnel costs per CIAL[1]	290.00
Cost of experiments per CIAL[2]	90.00
Other operational costs per CIAL[3]	122.00
Total cost per CIAL	502.00
Cost per Capita	*$ per Capita per Year*
Total population (50,000)	00.55
CIAL communities (12,900)	2.10
33% of CIAL communities (4,260)	6.50
Seed purchasers (10,500)	2.60
CIAL committee members (220)	125.50

Notes:
[1] Includes agronomist (0.33), farmer coordinator (1.0), and support-farmers (2.0).
[2] Average of costs per CIAL charged against CIAL funds in 1994.
[3] Average of transportation, supplies, and capital depreciation on four motorcycles.

It is difficult to find published data on the costs of performing adaptive research with groups of farmers, which can be compared with the figures in table 5.7. These show the total operating cost per CIAL ($502.00 per community per year) and annual per capita cost, which ranges from $125.00 (if only the 220 farmers who are committee members are considered) to under $1.00 (if based on the total population in the area of influence), to $6.50 per capita (if it is assumed that only one-third of the population in the CIAL communities actually receive any contact with their CIAL's adaptive testing). Based on the estimated number of purchasers of CIAL seed, the cost per capita would be approximately $3.00. The total annual operating budget of the CIAL Corporation currently amounts to the equivalent of about two agronomists' salaries at national program rates.

These figures compare favorably with costs cited by Nimlos and Savage (1991) of $36.00 per capita and $2,664.00 per community, annually, for an extension program using village-level support-farmers in Ecuador. Also in Ecuador, Romanoff (1993; cited in Bebbington, Merrill-Sands, and Farrington 1994) reports the cost of forming groups of between ten and thirty members using farmer-to-farmer training mechanisms at around \$3,000.00. However, these groups were farmer associations for processing and marketing cassava and were much larger and more complex than the CIALs.[6]

With respect to coverage, figures cited by Schwartz (1994:11–12) range from between one hundred and three hundred farmers per extension agent (private sector) to three thousand per extension agent (public sector), from case studies in Nigeria, Kenya, and Thailand. Comparable figures for the CIAL Corporation can be estimated at between sixty-six (direct contact with committee members) to around three thousand (population of the CIAL communities or seed purchasers, for example) per salaried paraprofessional or agronomist. However, since the CIALs do not perform a complete extension function at this time, but only a partial research or extension function to facilitate the adaptation of technology, this comparison is not completely equivalent.

One reason that the cost of forming and running CIALs is relatively so low may be that the procedures for creating these groups were formalized fairly early in the process, in the form of training materials written together with farmers that are easily used by farmers. Use of these materials means that support-farmers with practical experience in the procedures, who represent very low-cost labor, can form and run CIALs with minimal external support. Experience in Bolivia, for example, suggests that the CIAL handbooks can readily be used to form and run CIALs by extensionists without prior training in the method (Soria personal communication 1995). Nonetheless, the cost data presented here should not be viewed as conclusive, given that further testing of the approach without the intervention of the originators (the IPRA project team) is underway and will permit assessment of how robust and replicable the method is in different environments and with variant cost structures.

Equity

An issue related to assessing the effectiveness of the CIAL method for broadening the coverage of adaptive research is the question of how equitably benefits are distributed. The project has yet to conduct a comprehensive analysis to address this question, but survey data available on a subpopulation of eleven communities do provide some insight. Selection of CIAL members is predicated on the assumption that experimenting farmers are likely to be relatively better-off members of the rural community. Moreover, the CIALs are not designed to involve a large population in research. The committee mobilizes a capacity to test technology within the community on the basis of a limited participation in conducting the actual research. Therefore, distribution of knowledge about a CIAL's activities, rather than participation in it, is a more important test of the extent and nature of the coverage achieved.

A comparison of three social strata—differentiated on an index of well-being (Ravnborg 1994)—shows that of the sixty-four farmers actively participating in eleven CIALs, 39 percent come from the upper stratum, compared with 22 percent from the lowest stratum (Chi Square p = 0.046). Among the very poor, only 8 percent participate in CIALs, compared with 17 percent of the upper stratum. The community population is essentially one of small farmers, however, and in this subsample there is no significant difference in farm size between those who participate in the committees (average farm size of 4.4 hectares) and those who do not (average farm size of 3.5 hectares; probability of $t = 0.1484$). Knowledge of the CIALs is more evenly distributed: 52 percent of the population surveyed knows of the CIALs. There is no significant difference between the proportion of very poor people (49 percent) and the remaining two better-off strata (53 percent) who have this knowledge (Chi Square: p = 0.491).

The key issue is to what extent special interest groups in the community are able to get their priorities onto the agenda defined in the community diagnostic meetings—where it is decided which problems CIAL research will address. Monitoring by the project has detected a marked tendency for few women to attend these meetings, and those who do attend often propose research problems that are not prioritized. In order to address this need, the project established a separate fund for communities to set up a women's CIAL if a group of ten or more women so requested. Only two women's groups have formed CIALs, and four others have added women to the committees. Women still represent only 7 percent of committee members. The main reason for this appears to be the difficulty women have in devoting time to regular meetings that take them out of the home. For special interest groups like women, or the semilandless laboring poor, the research committee may not be an appropriate instrument for addressing their special research agendas.

Several options have yet to be explored. One option might be, for example, holding separate diagnostic meetings with special interest groups to identify priorities, which then could be included as treatments in trials carried out by CIAL members. However, this raises the question of the degree of motivation of CIAL members to carry out trials on topics of secondary importance to them and to the more powerful members of the community. Another option is to have special interest groups evaluate the trials so that their criteria for what is a desirable innovation are included in the analysis and recommendations drawn from a CIAL's research. It may be that increasing the equity of coverage by adaptive research has to be achieved by targeting the very poor, with the "slack" research or extension capacity of intermediary organizations created by devolving part of the research agenda to CIALs. These issues are topics for further empirical research.

Conclusions

Forming and monitoring the evolution of the CIALs is an ongoing experiment to assess the feasibility and the implications of devolving the responsibilities for adaptive agricultural research to farmers. The CIALs were formed to investigate to what extent the methods for participatory diagnosis and problem definition, planning and evaluation, and monitoring of adaptive technology testing could be handed over to community-level organizations—to generate in turn a "demand-pull" on formal research and extension systems and to improve the access of resource-poor farmers to an adaptive technology testing service at reasonable cost. Our experience so far suggests that it is possible to institutionalize this responsibility with farmers, that it is not unrealistic to expect "hard data" from farmer-managed adaptive research, and that this demonstration of farmers' capability wins respect for the farmers—a respect that is catalyzing a gradual reorientation of bureaucratic institutions' priorities. Results show that current costs and coverage compare favorably with some state or private sector systems, although the basis for comparison is limited. A favorable cost structure is clearly related to the demonstrated effectiveness of paraprofessionals such as support-farmers for scaling up and achieving devolution.

The project is entering a new phase, with an international training program and the monitoring of new CIALs, dispersed in widely contrasting sociocultural environments as distant as Brazil and Honduras. Many questions remain, concerning the long-term viability of the CIALs as an approach to institutionalizing farmer participation in agricultural research. But there are already some signals—such as CIAL Miske in Bolivia, which is reaching out to serve twenty-two communities, or the CIALs in Peru, which organized as a group to campaign for support from their regional state experiment station—that a quiet revolution may be underway to bring the capacity of farmers as researchers to fuller recognition.

APPENDIX 1

Procedure for the Formation of a New CIAL

1. On-farm research and extension staff of the host institution receive training in the CIAL methodology and then select communities, or respond to requests from communities, to form a CIAL.[7] The host institution may be a state agency, an NGO, or a farmer cooperative.

2. The host institution calls a community meeting where farmers undertake a group analysis of what it means to experiment with new agricultural practices, of local experience with experimentation and results, and of the purpose of a local research committee.

3. If the community decides to establish a CIAL, it elects a four-member committee of farmers who are recognized locally as experimenters, with leadership qualities as defined by the community before the election.

4. The CIAL conducts a diagnosis, in one or more community meetings, at which a topic for the CIAL experiment (such as crop, cultural practice, fertilizer use) is prioritized.

5. In a planning meeting with an agronomist from their host institution, the CIAL defines objectives of their experiment, treatments and controls, cri-

teria for site selection, scheduling, inputs, data needed to draw conclusions from the trial, and responsibilities for different tasks. In the first training cycle, a support-farmer visits the CIAL on a regular basis as these tasks are implemented.

6. Once the experiment is planned, the CIAL carries out the activities involved (from planting to harvesting) and manages the community's CIAL fund. This is a collective rotating fund in which each CIAL has a share. In Colombia, the CIAL fund amounts to less than 50 percent of the value of a head of livestock in the pilot area ($375.00 per CIAL at current exchange rates).

7. Once the experimental crop has been harvested, the CIAL meets with the agronomist, to draw conclusions from the data they have collected, and plans the community meeting where the CIAL will present its results.

8. The community meets to hear an oral report by the CIAL of its activities, results, and financial status. If appropriate, the diagnosis is repeated to orient the CIAL's activities for the next season.

9. In the second and subsequent cycles of experimentation, two or three monitoring visits are conducted by the support-farmer.

APPENDIX 2

CIAL Cooperation

The following example illustrates how CIALs work together. In 1991, Loma Corta prioritized field peas in their community diagnostic meeting with the objective of finding a short-season crop that was easy to cultivate, had a stable price, was used for consumption as well as for sale, and was easy to market. Field peas were a completely new crop in the CIAL experiments. This region is considered marginal for field peas, and the crop is not officially recommended. Loma Corta planted a varietal trial with four varieties obtained by the paraprofessional from an experimental station in another department. In another experiment, Loma Corta compared three systems of support for field peas: posts with string (the technical recommendation); bamboo stakes collected from local groves; and bamboo stakes with one-third of the amount of string recommended. CIAL Loma Corta discarded the technical recommendation of posts with string because the string cost five dollars whereas the second system used only local materials and the third, 60 percent less string. Their budget analysis showed that the local support systems require more labor but less cash outlay.

Organizing Experimenting Farmers

At the annual meeting of CIALs *(Encuentro CIAL)*, CIAL Betania learned about Loma Corta's experiment with peas, which was then in its second cycle *(parcela de comprobación)*. As a result, CIAL Betania planted the two varieties and adopted the pea support system selected by CIAL Loma Corta. CIAL Esperanza—a community located in a colder climatic zone—repeated the varietal trial with the two varieties selected by CIAL Loma Corta to see if they were adaptable under two planting systems, line planting and *cajuela* (their traditional system of planting). After determining that line planting was preferred, because of the higher plant density obtained in a small plot close to the home garden, CIAL Esperanza planted a second experimental plot to test the two support systems with bamboo stakes. Loma Corta lost one year (two cycles) waiting for an agronomist who never fulfilled his promise to obtain more field pea varieties. Notwithstanding this demoralizing experience, Loma Corta went back to experimenting with the two varieties after observing the progress CIAL Betania was making with peas.

CIAL Betania, having learned from Loma Corta that the support system of bamboo stakes with string was preferred, selected one variety of pea that had the best commercial quality and that the women preferred for its large size *(Piquinegra)*. They planted a production plot and began to sell the produce. They decided to sell part of the harvest fresh and another part as seed (the latter being worth 120 percent more than the fresh peas) to other farmers in Betania and the surrounding region.

On the basis of Betania's experience, Loma Corta planted *Piquinegra* with the bamboo stakes and string system and immediately went to production plot and seed multiplication. Esperanza, having tested variety, planting system, and support system to its satisfaction, scaled up to production plot and seed multiplication, buying seed from Betania. Now field peas are beginning to appear in monocrop fields and in association with other crops in farmers' fields. This occurred after Betania took its seed for distribution to the *Encuentro CIAL*.

6

Technologies for Sustainable Forest Management
in the Northern Zone, Costa Rica

Carlos Reiche

This chapter describes local and introduced technologies for sustainably managing tropical forests in the Northern Zone of Costa Rica. The Costa Rican government has used fiscal incentives as a major policy tool in the forestry sector. In 1991 incentives to landowners were institutionalized to promote sustained-yield management of their forests. The Forest Certificate Bond for Management (CAFMA) compensates for the opportunity costs involved in extracting only 60 percent of the standing commercial timber and leaving the remainder for future regeneration and harvesting. The landowners derive economic benefits from this policy. Within the Northern Zone, more than five thousand hectares are under CAFMA management plans. These serve as examples to non-participating landowners to implement similar plans.

   In this chapter we present the technical and economic results of sus-

The author is grateful to Eva Müller, Jonathan Davis, Jörg Linke, and Jochen Weingart for their reviews, comments, and suggestions concerning earlier drafts of this chapter. However, the content is the sole responsibility of the author.

tainable management technologies currently being implemented in the tropical forest of Costa Rica's Northern Zone. Preliminary financial analysis indicates that, under incentive program management plans, forest owners are able to turn a profit while still maintaining the indirect benefits of ecosystem preservation for society as a whole. We also analyze secondary forest production options. Natural regeneration, management, and treatments are the basic techniques employed. We will consider legal, cultural, and financial constraints to secondary forest production as well as the potential for sustainable forest production, protection, and other indirect benefits.

The Northern Zone of Costa Rica borders Nicaragua and covers 7,662.83 square kilometers (15.14 percent of the country's total area). It includes the cantons of San Carlos, Los Chiles, Upala, and Guatuso, as well as the districts of Sarapiquí, Río Cuarto, and San Isidro Peñas Blancas. The area is comprised mainly of flatlands formed from volcanic material—geologically part of the coastal floodplains of the Atlantic Ocean. There also are hilly areas (with altitudes up to two thousand meters and terraces ranging from fifty to one hundred meters in altitude), alluvial plains, and some extended areas with swamps.

In ecosystemic terms, the Northern Zone is classified as a "premontane very humid forest life zone" (about 52 percent) or a "very humid tropical forest life zone" (24 percent), with the remainder corresponding to other minor categories. Nearly 70 percent of the soil is classified as forestland, although only about 19 percent is actually under forest cover. Other vegetative complexes include pasture (62 percent), agriculture (9 percent), and bush (7 percent). Temperature and annual rainfall averages are 24 degrees Centigrade and 3,271 millimeters.

The population estimate for 1994 was 177,659 inhabitants (52 percent males, 48 percent females), with 87 percent living in rural areas. The forest inventory of the Northern Zone shows that, of the total forest area, 26 percent (34,375 hectares) is classified as undisturbed forest, 46 percent (66,875 hectares) as disturbed or intervened (logged-over) forest, 16 percent (20,000 hectares) as secondary forest, and 14 percent as forest plantation (COSEFORMA 1994). The size of farm forestland ranges from 10 to 300 hectares. Currently, logging takes place during the dry season (March to April).

Carlos Reiche

Traditional Technologies

Unplanned, highly forest-degrading timber extraction has been the general pattern of traditional forest use in the region. Timber extraction is not actually the main cause of deforestation, however. More than 80 percent of the nation's deforestation results from the conversion of forest to pasture (alone accounting for 50 percent) and agriculture. Where cleared land was developed for pasture and agriculture, deforestation and the extraction of timber were initially beneficial to loggers, but evidence shows that most of the land being cleared could not sustain farming or grazing under traditional practices. Consequently, much of the cleared land was abandoned after only a few years.

Timber extraction has generally taken two forms. The first is the selective cutting of valuable timber in primary, intervened, or secondary forest under a so-called forest management plan. Since the best trees were extracted and consideration was not given to natural regeneration, this practice generally led to forest degradation. The second is the exploitation of pasture areas that contain patches of forest or individual trees, cut by the landowners for extra income or to eliminate excessive shade (Lutz et al. 1993). From the tropical forest and pasture areas remaining in Costa Rica, more than 350,000 cubic meters of timber (more than 42 percent of the nation's industrial demand) are harvested each year through these two forms of forest intervention. The preferred species are those that provide the finest and most decorative woods and have the highest market values—caobilla *(Carapa guianensis)*, almendro *(Dipteryx panamensis)*, cedro María *(Calophylum brasiliense)*, cocobolo de San Carlos *(Vatairea lundellii)*, botarrama *(Vochysia ferruginea)*, and fruta dorada *(Virola koschnyi)*.

The quantity of timber extracted under traditional practices depends on the major role-player in the commercialization of timber—the logger. The logger buys the standing timber from the forest owner at a relatively low price—$16.00 per cubic meter for soft timber; $34.00 per cubic meter for semihard timber; and $44.00 per cubic meter for decorative timber (prices as of March 1995).[1] The logger also takes responsibility for dealing with the many bureaucratic procedures that must be negotiated in obtaining official permits for cutting trees and transporting logs to the sawmill. He also provides his own machinery, equipment, and the necessary labor to cut and extract the timber. Traditional methods involve building roads through the forest, without any concern for natural regeneration or future rotations, and using heavy machinery in the logging operations—cutting all trees greater than sixty centimeters in diameter at breast height (DBH). Methods of felling and extracting cause severe damage to the remaining stand. It is estimated that under this system, 2.14 square meters are damaged for each square meter of basal area cut.

Improved Technologies for Sustainable Management

Currently, the increased intensity of natural forest use tends to sacrifice its renewability. The forest inventory for the Northern Zone indicates that the area will soon experience shortages of forest products and services. The objective of forest resource development is to provide people with such products and services. Future generations are expected to need forest products and services as much as the present generation.

The challenge is not to restrict forest use so much as to change traditional practices. The goals are to ensure that logs and other forest products are harvested under sustainable management systems, that forest area is cleared only in a planned and controlled way, and that subsequent land uses are productive and sustainable. To achieve these goals, appropriate technologies are needed, to sustain primary, intervened, and secondary forest resources. Although technologies exist, they are not in general applied in the Northern Zone to prevent the conversion of productive forestland (from undisturbed, disturbed, secondary forest, and plantation) to less productive options (such as degraded forestland, grassland, and crops). The challenge, then, is to demonstrate to forest owners that conserving and managing their forest resources can be financially beneficial in the short as well as in the long term.

Currently, the first steps are being taken toward reaching the goal of harvesting timber under sustainable management practices in the

Northern Zone. Local and international organizations—such as the German Technical Cooperation Project (GTZ) through its Cooperation in the Forestry and Wood Sector (COSEFORMA) Project, the San Carlos' Commission for Forestry Development (CODEFORSA), the English Overseas Development Agency (ODA), and the Agriculture, Industry and Forestry Producers Association (APAIFO)—are involved in the search for sustainable forest resource management strategies. In addition, the government is providing fiscal incentives to forest owners to apply improved management techniques. The resulting process is likely to be complex. For example, one key approach—the Regional Plan for Forestry Development—requires the active participation of various institutions, individuals, and target groups involved in the Regional Forestry Sector under the promotion and guidance of COSEFORMA.

Technologies for Natural and Disturbed Tropical Forest

A variety of management practices help minimize the impact of intensive harvesting and contribute to sustainable forest management. Local and introduced technologies include (1) management plans, as a key tool to assure correct management;[2] (2) forest inventory, for planning and for mapping forest areas; (3) harvesting of only 60 percent of the standing commercial timber (measured in cubic meters), and not 100 percent, as often takes place in traditional extraction; (4) preplanned road and skid-road construction that avoids damage to natural regeneration; (5) observing optimal tree-felling direction to minimize damage to the remaining stand; (6) new techniques for skidding in order to minimize damage to soil and vegetation;[3] (7) protecting seed trees (known as mother trees) to promote the regeneration of desirable species; (8) protection zones along rivers and natural water courses; (9) post-harvest silvicultural treatments; and (10) follow-up for second harvests. The minimum harvesting diameter of 60 centimeters DBH is based on the type of sawmills (which were designed to process large-diameter logs) already located in the Northern Zone.

Combined Traditional and Modern Technologies for Secondary Forest

In the absence of plantations of harvestable age, the present demand for industrial wood in Costa Rica (estimated at 830,000 cubic meters per year in 1995) must be met through harvesting forests without eliminating or degrading them. However, the area of natural forest has been reduced to such an extent that demand cannot be met by this type of forest alone. Secondary forests are a new forest resource in the Northern Zone and have the potential to meet some of the current and future demands for timber. For the purpose of this chapter, secondary forest is defined as the vegetative complex that becomes established on abandoned crop or pasture areas through a process of natural succession— beginning with the formation of secondary growth and passing through different stages to mature forest that resembles primary forest.

When cleared forestland is incapable of sustaining an introduced land-use system over time, the consequences may include land degradation, abandonment, or the reestablishment of forest cover. The latter generally depends on the intensity of land use, the degree of degradation that has occurred, and on whether the site has been grazed or burned. In the past, many foresters considered secondary forest of little use, because its trees were considered to have poor form, to consist of inferior species, or to be too small in diameter. In some cases, financial incentives were offered to eliminate young secondary forests in order to establish plantations. In addition, promoting secondary forest growth was considered a poor form of management. The land, as "abandoned" farmland, was in danger of being invaded by landless peasants. Foresters as well as farmers must change their attitudes about secondary forest if these vegetative complexes are to serve as an additional forest resource.

Approximately twenty thousand hectares of potentially productive secondary forest exist in the Northern Zone; about five thousand of these are in protected areas (COSEFORMA 1994). This region of secondary forest is composed of a mosaic of small areas on abandoned pasturelands. The sizes of eleven secondary forests studied in Boca Tapada ranged from 1.5 to 16 hectares (average 6.7 hectares) with ages between 4 and 20 years (average 10.9 years) (Spittler 1995).

Secondary forest, if managed properly, could satisfy growing timber needs. Technologies to manage secondary forest are not yet refined enough to assure suitable returns on investment. However, they have potential for the production of timber as well as other benefits such as wildlife habitat, carbon sequestration, and sources of medicinal plants. The technologies currently used to manage secondary forests vary in the intensity of treatments.

The first "technology" is *no treatment.* This is the most common system used in the Northern Zone. In general, landowners allow the natural vegetation to reestablish itself in abandoned areas. Locally, this natural process is called *encharralar.* In its early stages, the evolutionary process from secondary forest to tropical primary forest is characterized by a high prevalence of pioneer species. Gradually, the pioneers are replaced by more permanent light-demanding species, and by species that need partial shade and that generally grow more slowly. Eventually, the vegetative composition changes to that of primary forest, with a high proportion of both shade-living and light-demanding species. Untreated secondary forests are low input/low output systems with, therefore, low costs and values because there are no silvicultural interventions.

In general, and especially during the early stages of development of a secondary forest, the owners do not realize that the "disorder" on their farms is a growing forest resource with a potential future value. From the fifth year on, valuable species reach heights of between five and six meters. For the landowners to see the value in maintaining the secondary forest in these early stages, they must be shown that the forests will produce acceptable yields within a relatively short time. Often by the eighth year, commercial trees can be extracted for the first time, thus regulating the stand composition and concentrating its growth potential. Preliminary studies of unmanaged secondary forests aged between two and a half and eighteen years in Boca Tapada and Pital produced a total volume of 185 cubic meters per hectare at eighteen years (see table 6.1). Of this volume, 141 cubic meters per hectare (76 percent) were commercial and 82 percent of the trees had DBHs greater than 15 centimeters (Fedlmeir 1995). Processing small-diameter logs from secondary forests requires a complementary industrial technology. For instance, portable sawmills have the potential to process small-diameter logs and to minimize access problems.

The second technology currently used to manage secondary forests is *silvicultural treatments.* One silvicultural treatment (known as refine-

Table 6.1. Annual Mean Increment and Volume Estimation at Different
Ages from Secondary Forest in Boca Tapada and Pital,
Northern Zone, Costa Rica

Secondary Forest (age in years)	*Volume Increment* (m³/ha)	*Total Volume* (m³/ha)	*Commercial Volume* (m³/ha)
2.5	8.2	4	2
5.5	14.0	22	6
9.0	10.3	81	26
12.0	14.2	104	83 (54 m³ ≥ 15 cm DBH)
18.0	12.6	185	141 (116 m³ ≥ 15 cm DBH)

Source: Fedlmeir 1995.

ment) consists of removing the less valuable trees, which reduces com-
petition and provides more growing space to the valuable trees. This
thinning can be carried out at different intensities and to obtain different products. One example is the use of thinning by a secondary forest
owner from the community of Gansos, Boca Tapada, in the Northern
Zone, to selectively promote the growth of botarrama *(Vochysia ferruginea)*, a species with a locally high commercial value (Spittler personal
communication 1995). Some refined secondary forests also produce non-
wood products such as medicinal plants, palm leaves for roof construction, and other materials.

To some extent, liberation thinning is a variation of refinement. In
liberation thinning, desirable trees are identified and liberated from com-
petition with less desirable species. Another variant of refining is the
polycyclic system, which helps to preserve the structure of the natural
forest by periodically harvesting mature trees and thus liberating im-
mature trees.

A different option is the monocyclic system. The goal of this system
is to have all trees reach maturity simultaneously for a single harvest.
In this system, all undesirable species above ten centimeters DBH are
removed. Drawbacks to this treatment are the initial sacrifice of larger
trees and the lack of an intermediate harvest for added income. Moreover, monocyclic management ultimately leads to a more intensive complete harvest and the increased potential for soil erosion. Heavy cuts
from monocyclic management can be expected to lead to the periodic

interruption of nutrient recycling, since large volumes of logging slash will be decomposing at a time when an effective network of root to capture the nutrients is lacking (Office of Technology Assessment 1984).

To date, there are no clear legal guidelines for secondary forest management in Costa Rica's national legislation. For instance, it should be illegal to convert secondary forest back into pasture once it has reached a predetermined height. However, on the technological side, an established market does not exist for smaller dimension logs from forest plantations. Consequently, a change in industry practices will have to take place so that appropriate equipment for processing such timber will be adopted. This would make secondary forest legislation more practical. There are several important steps to be taken regarding secondary forests: (1) the legal clarification of their status; (2) establishment of an extension program to emphasize their importance to target groups; (3) design and application of a regional management scheme; (4) identification of markets for smaller dimension logs; and (5) government subsidies to support the initial phase of establishment.

Incentives for Sustainable Forest Management

The government of Costa Rica has established various fiscal incentives as part of a policy strategy to promote the forestry sector. The Forest Service (DGF) is the government agency responsible for controlling the logging of all forests within the country. In 1991 incentives were institutionalized to encourage sustainable management of natural forests.

The Forest Certificate Bond for Management (called CAFMA) is an incentive designed to compensate for the opportunity costs lost when only 60 percent of standing commercial timber is cut, leaving the remainder for future regeneration and harvesting. CAFMA is worth 80,175 colones per hectare or $459.46 ($1.00 = 174.50 colones in April 1995). It is distributed over five years: 50 percent is assigned in the first year (20 percent at the implementation of a management plan and 30 percent after the first harvest); 20 percent in the second year; and 10 percent per year over the remaining three years. From 1992 to 1994, about 114 million colones or $653,000 were disbursed for management

Technologies for Sustainable Forest Management

Table 6.2. Management Plans: Hectares and Amount of Assigned Colones
for CAFMA (Phase 1 and 2), Northern Zone, Costa Rica,
1992–1995

| Period | Phase 1 | | | | Phase 2 | | |
	Number	Hectares	Colones[1]	Number	Hectares	Colones	Total Colones
1992–93	6	761.33	12,282,328.00	9	907.51	36,396,399.80	48,678,727.80
1993–94	18	1,930.11	31,014,253.20	14	1,565.61	61,154,622.15	92,168,875.35
1994–95[2]	8	625.85	10,028,125.00	17	1,566.75	62,846,259.00	72,874,384.00
Total	32	3,317.29	53,324,706.20	40	4,079.87	160,397,280.9	213,721,987.10

Source: Tables A and B, Manuscript on CAFMA, DGF Region H.N., January 1995.

Notes: 1. US $1.00 = ¢182.84 (colones), August 1995.

2. COSEFORMA estimation based on applications for CAFMA with corresponding percentages.

plans in the Northern Zone. Of the approved applications, 52 percent were for phase 1 management plans (first harvest), which absorbed 31 percent of the total amount approved; 48 percent were for phase 2 management plans (post-harvest silvicultural treatments) and absorbed the remaining 69 percent of assigned incentives (see table 6.2). The average forest area under each management plan was about one hundred hectares.

This policy has technical and economic implications. Until 1993–1994, a little more than 5,000 hectares were under management plans in the Northern Zone. Of these, 90 percent were estimated to be well developed and qualifying for phase 2 (Vargas 1995). These successful examples demonstrate to other landowners that adopting a management plan can contribute to sustainable practices. For fiscal year 1994–1995, twenty-five additional applications were ready to be financed, encompassing a total area of 2,192.6 hectares (625.85 hectares corresponding to phase 1 and 1,566.75 hectares corresponding to phase 2). In sum, a total of 7,397.16 hectares were scheduled to be under management plans and the incentive program by the end of 1994–1995.

Preliminary financial analysis of some cases in the Northern Zone indicate that forest management can be economically profitable. For example, the San Juan Cooperative (COOPESANJUAN R.L.) obtained a 41 percent return, or $19.60 per cubic meter (Solís and Reiche 1995). La Tirimbina obtained a 50 percent return, or $25.97 per cubic meter, and Corinto obtained a 51 percent return, or $23.79 per cubic meter (Finnegan et al. 1993).

Projections for second and subsequent harvests indicate positive net present values (NPV). In the case of COOPESANJUAN R.L., an NPV of $196.29 per hectare was projected within a period of twelve years, with a discount rate of 5 percent (Solís and Reiche 1995). In the case of La Tirimbina, an NPV of $340.00 was calculated, using a forty-year period, with two harvests and a 22 percent discount rate. For Corinto, an NPV of $411.00 was estimated, using a forty-year period and a discount rate of 22 percent (Finnegan et al. 1993).

Conclusions

Although this chapter uses site-specific data from only one area in Costa Rica, it illustrates that improved technologies can contribute to sustainable yield forest management. Traditional intensive harvesting has proved to be destructive and results in a significant reduction of natural forest cover. Technologies to improve management practices exist. The modest use of simple management plans and better techniques for harvesting, transportation, and handling can reduce the destructive impacts of logging and at the same time contribute to forest maintenance. The utilization of directional felling (to protect the nonharvested trees) and planned skidding trails alone reduce damage by as much as 14 percent. Cable yarding is another simple technology that has been successfully applied to reduce road requirements and soil disturbance.

Existing secondary forest, if managed properly, could satisfy the ever-increasing demand for timber. Appropriate technologies to increase yields include the use of thinning, liberation, and improved harvesting practices. However, to enhance the value of secondary forests, markets for lesser known and smaller diameter tree species need to be developed.

PART IV

Indigenous/Local
Knowledge Systems

7

Indigenous Knowledge for
Agricultural Development

D. Michael Warren

On 3 December 1990, Michael Cernea provided me with the opportunity to present a seminar at the World Bank on "Indigenous Knowledge and Development." This presentation was the basis for World Bank Discussion Paper No. 127, *Using Indigenous Knowledge in Agricultural Development* (1991a), which has been widely disseminated.[1] What I propose to present in this chapter is an overview of events and activities that have occurred in the past five years in the area of indigenous knowledge and development, focusing on agriculture and natural resource management.

The first publications to use the term "indigenous knowledge" in a development context were those edited by Robert Chambers (1979) at the Institute of Development Studies, University of Sussex, and by Brokensha, Warren, and Werner (1980). Both publications presented case studies that explored the potential of community-based knowledge and decision-making systems for bridging the communications gap between communities and development professionals. It was envisioned that this "bridge" would facilitate mutual understanding of problem situations and enhance participatory decision making. By using ethnoscientific methodologies to record local knowledge systems, development practitioners would have access to the knowledge that influences both individual and communitywide decisions involving the problem identification process as well as the search for solutions. By recording local knowledge, scientists could compare and contrast the local system (such as a soil classification system) with its global counterpart, in order to understand the comparative strengths and weaknesses within each system.

What appeared to many of us to be a straightforward approach to improving the development paradigm—by adding indigenous knowledge components that would facilitate participatory decision making, capacity building, and sustainability—was often misunderstood by some in the development community. Although our focus was on contemporary indigenous knowledge systems, many practitioners felt that indigenous knowledge represented an unfortunate step backward, into a distant past. When we argued that an indigenous knowledge component would strengthen the project cycle as well as farming systems and training and visitation approaches to agricultural research and extension, the reaction was often neutral, even negative. It was clear we were facing the uncomfortable reactions associated with what was regarded by many in the development profession as a paradigm shift.

The 1991 publication of World Bank Discussion Paper No. 127, however, resulted in more critical attention being paid to the role of indigenous knowledge in facilitating sustainable approaches to development. Much of what I intend to enumerate in this chapter has been achieved because of the interest of World Bank professionals such as Michael Cernea, Scott Guggenheim, Cynthia Cook, and Charles Antholt in our attempt to design a more cost-effective approach to improving the quality of life in communities living in marginal environments. The most important achievement in the past few years has been a better understanding of the dynamics generating new indigenous knowledge. The early stages of interest in the role of indigenous knowledge in development focused on the knowledge itself, and how it could be reflected as transects and taxonomies. We now understand the crucial role that such knowledge plays in decision making, how indigenous organizations facilitate the identification and prioritization of community problems, and how discussions of solutions to problems result in local-level experimentation and innovation.

In World Bank Discussion Paper No. 127, I was asked to enumerate areas that the Bank could reflect upon and possibly support either directly or indirectly. The areas I identified were many. They included understanding the relationship of indigenous knowledge to biodiversity; strengthening the capacities of indigenous knowledge resource centers; improving field methodologies for recording indigenous knowledge and decision-making systems; utilizing indigenous knowledge in educational institutions; undertaking research on change in knowledge systems induced through local-level experimentation and innovation; and supporting global networking through newsletters and electronic communication.

Cultural Capital

The timing for the discussion paper's appearance could not have been more opportune, as it was available to and influential in many of the discussions that took place at the U.N. Conference on Environment and Development, held in Rio de Janiero in June 1992. The Agenda 21 documents based on this conference include numerous references to indigenous knowledge as national and global resources. These resources are in as much jeopardy as the biological resources being depleted by various forces of modernization.

It is no longer questioned that indigenous knowledge—what Berkes and Folke (1992) have called "cultural capital"—is a highly valuable resource. Biodiversity surveys are strengthened by including the local knowledge of the biological capital for a community. Efforts in this area are currently being supported by the Natural Resources Institute at the University of Manitoba, in conjunction with the Beijer International Institute for Ecological Economics, part of the Royal Swedish Academy of Sciences. A multidonor project being coordinated by Fikret Berkes in Manitoba and Carl Folke in Stockholm will result in a major contribution to our understanding of the dynamic nature of community-based knowledge as it relates to the biological realm, including both domesticated and nondomesticated plants (Berkes and Folke 1992; Warren and Pinkston 1997).

Science Education

A content analysis of the current curricula for K–12, universities, teacher-training colleges, polytechnics, and extension training institutes in many countries reveals that the knowledge generated by the communities within a country is rarely reflected in the educational system. Mathematics and science are often regarded as Western phenomena by both teachers and students. The result of this attitude has been a devaluation and disregard of local knowledge. By analyzing national education policy, changes can be effected that allow cultural capital to be added to existing curricula (Titilola et al. 1995; Warren, Egunjobi, and Wahab 1996). The rapidly growing database of case studies of indigenous knowledge in an ever-growing array of disciplines (such as agronomy, animal science, veterinary science, mathematics, astronomy, plant pathology, climatology and meteorology, entomology, engineering, ecology, forestry, and aquatic resource management) has resulted in efforts at the Center for Indigenous Knowledge for Agriculture and Rural Development (CIKARD) to produce teaching modules that will be made available on the Internet (Kroma 1995). Presentation of local knowledge that parallels global knowledge allows students the opportunity to discover the links between the two and to understand that the communities from which local knowledge comes have made important contributions to global knowledge.

Recent efforts to rectify this situation include providing user-friendly training manuals and training opportunities for teachers, extension workers, and development professionals. The training manual series of the International Institute for Environment and Development (IIED) now has twenty-two volumes, which are still available free to persons in developing countries. The series has been an important mechanism for providing global access to participatory research methods. The Dutch Ministry of Foreign Affairs has supported production of a training manual focused on participatory approaches to development, which is now available in both Spanish (Bojanic et al. 1994) and English (Gianotten and Rijssenbeek 1995).

The Regional Program for the Promotion of Indigenous Knowledge in Asia (REPPIKA) has published a manual on recording indigenous knowledge for sustainable approaches to development, which is now available through the International Institute for Rural Reconstruction (IIRR), Silang, Cavite, Philippines (Mathias 1997). This manual is used for a new international course at IIRR on Applying Indigenous Knowledge in Development. Since IIRR has national programs in countries in Asia, Africa, and Central and South America, such a course has an immediate multiplier effect. A field manual for using the laptop computer in recording indigenous knowledge (such as land-form transects, taxonomies, and decision matrices) has been tested in the classroom at CIKARD (Ames 1995). It was field tested in Nigeria in summer 1995 in the area of gendered knowledge reflected in biodiversity and soil science (Rechkemmer 1996).

A Global Network of IK Resource Centers

The number of indigenous knowledge centers with global, regional, and national mandates has expanded rapidly since 1990. There are now four global centers, two regional centers, and twenty-six national centers (see the appendix at the end of this chapter). Centers now exist on each continent, including Europe (Netherlands [2], Russia, Greece, Georgia), North America (United States [2], Canada), Latin America (Mexico, Venezuela, Brazil, Uruguay), Africa (Burkina Faso, Ghana [2], Nigeria [4], Tanzania, Ethiopia, Sierra Leone, Kenya, South Africa, Cameroon, Madagascar), and Asia (Indonesia, Sri Lanka, Philippines [2], India [2]). Negotiations are currently underway for additional regional centers for the fourteen South Pacific island nations and the southern African countries, as well as national centers in Benin, Namibia, Zimbabwe, Costa Rica, Colombia, Peru, Honduras, Bolivia, Nepal, Australia, China, Vietnam, Mali, Senegal, Pakistan, Thailand, Papua New Guinea, Sudan, and Israel.

Global networking has moved ahead with support from the Dutch

Ministry of Foreign Affairs, the U.S. Agency for International Development (USAID), the U.S. Information Agency (USIA), Canada's International Development Research Center (IDRC), the European Community, the Ford Foundation, and the Royal Swedish Academy of Science (SAREC). Funds have allowed indigenous knowledge centers and practitioners to hold national, regional, and global workshops in the Philippines, India, Sri Lanka, Sudan, China, South Africa, Nigeria, and Indonesia (see CIAD 1994; Normann, Snyman, and Cohen 1996). These workshops have provided opportunities for practitioners from different countries to get to know one another personally and to expand the numbers of professionals knowledgeable about the methodologies for recording indigenous knowledge. The African Resource Center for Indigenous Knowledge (ARCIK)—based in Ibadan, Nigeria, and supported by IDRC and the Ford Foundation—has conducted a West African regional workshop as well as three national workshops in Nigeria for farming systems professionals, NRM specialists (including foresters), hydrologists and soil scientists, and representatives of NGOs working with indigenous organizations.

Global networking has been greatly enhanced through the dissemination of newsletters such as the *Indigenous Knowledge and Development Monitor*, published at the Center for International Research and Advisory Networks (CIRAN) in The Hague (Tick 1993; von Liebenstein 1994; von Liebenstein and Tick 1994). The *Monitor* is sent to more than 3,000 recipients in 120 countries.[2] Information on each of the participants in this network is currently being compiled at CIRAN for a global directory of individuals and institutions involved in indigenous knowledge activities. Other periodicals that focus on indigenous knowledge include *ILEIA Newsletter* (see Sherwood and Bentley 1995); *Honey Bee* (see Gupta 1990, 1993); *Etnoecologica* (see Reichhardt et al. 1994; Toledo, Ortiz, and Medellín-Morales 1994); and *SANREM Ecolinks* (SANREM 1993). Special issues of periodicals are now available on "Indigenous Economics," *Akwe:kon Journal* (see Barreiro 1992; Quintana 1992), "Indigenous and Traditional Knowledge," *IDRC Reports* (see International Development Research Center 1993), "Traditional Knowledge in Tropical Environments" and "Traditional Knowledge into the Twenty-First

Century," *Nature and Resources* (see UNESCO 1994a and 1994b), and "Indigenous Agricultural Knowledge Systems and Development," *Agriculture and Human Values* (see Warren 1991b).

Both multilateral and bilateral donor agencies are recognizing the role of indigenous knowledge in sustainable development. These include the World Bank (Davis 1993; Grimshaw 1993; Warren 1991b); the Food and Agriculture Organization (FAO) (Herbert 1993); the International Board for Plant Genetic Resources (1993); the International Plant Genetic Resources Institute (Eyzaguirre and Iwanaga 1996); CIP/UPWARD (Prain and Bagalanon 1994); IDRC (Lalonde 1993); the International Labor Organization (ILO) (Warren 1995b); the Canadian International Development Agency (CIDA) (Obomsawim 1993); UNESCO (1994a, 1994b); U.N. Environment Program (Dowdeswell 1993; Warren and Rajasekaran 1995); and Coordination of Development (CODEL) (Wright 1995). The National Research Council (NRC) has sponsored and published important reviews and policy statements on indigenous knowledge (NRC 1992a, 1992b, 1992c). Research on indigenous agricultural knowledge has been supported by members of the Consultative Group on International Agricultural Research (CGIAR 1993), such as CIAT (Ashby 1987), IRRI (Fujisaka 1995), CIP/UPWARD (Prain, Uribe, and Scheidegger 1991), and the International Institute of Tropical Agriculture (IITA) (Warren 1992).

Intermediate Technology Publications (ITP) initiated a new series, *Studies on Indigenous Knowledge and Development*, in 1995. Volumes in this series include Blunt and Warren (1996), Castro (1995), Hess (1997), Innis (1997), McCorkle, Mathias, and Schillhorn van Veen (1996), and Warren, Slikkerveer, and Brokensha (1995). ITP has played an important role in publishing other volumes that link indigenous knowledge to sustainable agriculture (Chambers 1997; Chambers, Pacey, and Thrupp 1989; de Boef et al. 1993; Haverkort, van der Kamp, and Waters-Bayer 1991; Scoones and Thompson 1994; van Veldhuizen et al. 1997). Other major contributions to indigenous agricultural knowledge include Bunders, Haverkort, and Hiemstra (1996), Pretty (1995), Reijntjes, Haverkort, and Waters-Bayer (1992), and Sumberg and Okali (1997).

D. Michael Warren

Indigenous Knowledge on the Internet

On behalf of the global network of indigenous knowledge resource centers, CIKARD has entered into agreements with the Consortium for International Earth Science Information Networks (CIESIN), based in Saginaw, Michigan, and the International Institute of Theoretical and Applied Physics (IITAP), based at Iowa State University. With guidance and support from both CIESIN and IITAP, CIKARD has now established a home page on the Internet.[3] The following are available over the Internet: (1) the French and Spanish translations of World Bank Discussion Paper No. 127 (Warren 1991b); (2) French translations of several IK monographs and papers; (3) draft proposals and teaching modules on the topic of indigenous knowledge and education; (4) bibliographical citations and abstracts for the entire indigenous knowledge collection housed at CIKARD, currently numbering about forty-three hundred documents, which can be searched by key words; and (5) twelve annotated bibliographies produced by CIKARD research associates (with support from the Center for Global and Regional Environmental Research), based on documents housed at CIKARD's Documentation Unit. The topics (and number of citations) are agroforestry (324 citations), aquatic resources (109), biodiversity (61), ecology (356), ethnobotany (199), natural resource management (160), rainforest resources (69), soil knowledge (297), sustainable agriculture (354), terrestrial vertebrate wildlife (82), traditional medicine (158), and water management (186).

CIESIN has sponsored periodic workshops (beginning in August 1995) focusing on global, regional, and national database management and the use of electronic communications. The workshops are designed to facilitate the exchange of data at the global level. IITAP and CIKARD are currently exploring ways to expedite the transmission of indigenous knowledge teaching modules that could enhance mathematics and science teaching in developing countries.

Indigenous Knowledge and Agricultural Development

A literal explosion of published material on the role of indigenous agricultural knowledge in facilitating sustainable approaches to agriculture has occurred in the past several years. Now available are books by Adams and Slikkerveer (1997), Ahmed (1994), Brooke (1993), Brush and Stabinsky (1996), de Boef et al. (1993), Haverkort, van der Kamp, and Waters-Bayer (1991), Innis (1997), Leakey and Slikkerveer (1991), McCall (1995), McCorkle (1994), Moock and Rhoades (1992), Normann, Snyman, and Cohen (1996), Padoch and De Jong (1991), Prain and Bagalanon (1994), Pretty (1995), Reij (1993), Reijntjes, Haverkort, and Waters-Bayer (1992), REPPIKA (1994), Scoones and Thompson (1994), Slaybaugh-Mitchell (1995), Thurston (1992), Warren, Slikkerveer, and Brokensha (1995), and Winslow (1996). Frameworks for incorporating indigenous knowledge systems into sustainable agriculture have been presented by Rajasekaran (1994), Rajasekaran, Warren, and Babu (1991), and Warren (1994). The growing recognition of the active role that farmers play in experimentation and innovation is reflected in recent publications by Altieri and Merrick (1988), Bentley and Andrews (1991), Brush (1991), Bunders, Haverkort, and Hiemstra (1996), Chambers, Pacey, and Thrupp (1989), de Boef et al. (1993), DeWalt (1994), Eyzaguirre and Iwanaga (1996), Fujisaka (1995), Haverkort, van der Kamp, and Waters-Bayer (1991), Hecht and Posey (1989), Mathias-Mundy and McCorkle (1989), Mathias-Mundy et al. (1992), McGregor (1994), Niamir (1995), Pandey and Chaturvedi (1993), Pawluk, Sandor, and Tabor (1992), Posey and Balee (1989), Prain, Uribe, and Scheidegger (1991), Rhoades and Bebbington (1995), Sherwood and Bentley (1995), Sumberg and Okali (1997), Tapia and Banegas (1990), Thurston (1992), van Veldhuizen et al. (1997), Warren (1995a, 1995c), Warren and Rajasekaran (1993, 1995), and Warren and Warren (1999). A new text has been prepared and edited by Gordon Prain, Sam Fujisaka, and D. M. Warren, *Biological and Cultural Diversity: The Role of Indigenous Agricultural Experimentation in Development* (1999).

One of the most exciting developments in recent years has been the discovery that farmers do experiment and share their knowledge in organized ways. Pretty (1991) has a detailed history of agricultural research and extension from the seventeenth century to the nineteenth century in Britain that is fascinating and eye-opening. Equally exciting are the studies supported by USAID on farmer innovation in Niger (McCorkle 1994; McCorkle and McClure 1995). McCorkle notes that the numerous examples of innovative agricultural techniques indicate that

> Nigerian farmers are clearly open to and actively seek out and apply new ideas in their plant and animal agriculture. Moreover, they plan, implement, and evaluate their own informal on-farm research trials. In the process, they demonstrate a sophisticated understanding of the complex interactions among numerous variables with which they must contend. Finally, there exists a rich body of local technical knowledge in agriculture, elements of which—with increased cross-regional communication and, in some cases, validation and refinement by formal research—could almost certainly be of use to farmers throughout the Sahel. (McCorkle 1994:32–33)

She also notes that

> There is a wealth of indigenous agricultural knowledge and expertise in the Sahelian countryside, some of which, with inputs from formal RD&E systems, could undoubtedly benefit farmers throughout this ecozone. . . . So long as farmers' own research efforts and their plethora of communication resources are ignored, much of donor investment in research and extension institutions and projects will be lost. Like it or not, both research and extension depend for their success upon farmers' own informal systems of technology validation and transfer for success. (McCorkle 1994:39)

Another text, *Indigenous Organizations and Development*, edited by Peter Blunt and D. M. Warren includes case studies of the role of indigenous farmers' organizations in local-level experimentation. These case studies have enormous policy implications for agricultural research and extension.

Indigenous knowledge and global knowledge complement one another and can be compared and contrasted (see Warren 1989) in ways that can influence national and international agricultural research agendas in cost-effective ways. Fujisaka (1995) has, for example, recorded the positive and negative characteristics that farmers recognize in varieties of rice—information of great interest to plant breeders. The USAID-funded Iowa–Nigeria University Development Linkages Project has a major indigenous agricultural knowledge component. One of the activities currently being undertaken in Nigeria is the systematic survey of farmer knowledge of major cultivars in both the rainforest and the savanna. This information will be used by biotechnologists at an international seed company based in Iowa, Bioseed Genetics International, to improve seeds for Nigerian farmers by genetically removing negative characteristics. A recent study by Berlin (1992), based on analyses of ethnobiological knowledge systems from all parts of the globe, indicates the similarity of many ethnobiological classification systems to the Linnaean system of biological classification.

Based in part on the publicity provided by the publication of World Bank Discussion Paper No. 127 in 1991, major advances have been achieved in the past few years that provide a clearer understanding of the role of indigenous knowledge in agricultural and NRM projects. There is, of course, much still to be done—with support from development donor agencies. Such endeavors include the need to support activities that will increase the capacity of regional and national indigenous knowledge resource centers to conduct national inventories of indigenous knowledge in ways that make them available to both the communities of discovery and the development practitioners. Major efforts will continue, to expand access to the Internet in developing countries in cost-effective ways, so that indigenous knowledge documents, information, and data can be more easily exchanged. The role of indigenous knowledge in facilitating science education will continue a high priority. Support for global networking efforts, such as the *Indigenous Knowledge and Development Monitor*, must continue to be a priority for donor agencies.

D. Michael Warren

APPENDIX

Established IK Resource Centers, March 1997

GLOBAL IK RESOURCE CENTERS

1. Center for International Research and Advisory Networks (CIRAN): Dr. G. W. von Liebenstein, Director, Nuffic/CIRAN, P.O. Box 29777, 2502 LT The Hague, The Netherlands (e-mail Lieb@nufficcs.nl).

2. Center for Indigenous Knowledge for Agriculture and Rural Development (CIKARD): CIKARD, 318 Curtiss Hall, Iowa State University, Ames, Iowa 50011, U.S.A. (http://monet.npi.msu.su//iitap-mirror/cikard/cikard. html).

3. Leiden Ethnosystems and Development Program (LEAD): Dr. L. Jan Slikkerveer, Director, LEAD, Institute of Cultural and Social Studies, University of Leiden, P.O. Box 9555, 2300 RB Leiden, The Netherlands (e-mail starkenb@rulfsw.fsw.leidenuniv.nl).

4. Center for Traditional Knowledge (CTK): Mr. Julian Inglis, Director, P.O. Box 3443, Station D, Ottawa, Ontario K1P 6P4, Canada (e-mail jtinglis @magi.com).

REGIONAL IK RESOURCE CENTERS

5. African Resource Center for Indigenous Knowledge (ARCIK): Prof. Adedotun Phillips, Director, and Dr. Tunji Titilola, Research Coordinator, ARCIK, Nigerian Institute of Social and Economic Research (NISER), PMB 5-UI Post Office, Ibadan, Nigeria.

6. Regional Program for the Promotion of Indigenous Knowledge in Asia (REPPIKA): REPPIKA, International Institute of Rural Reconstruction (IIRR), Silang, Cavite, Philippines (e-mail IIRR@phil.gn.apc.org).

NATIONAL IK RESOURCE CENTERS

7. Ghana Resource Center for Indigenous Knowledge (GHARCIK): Dr. Mensah Bonsu, Director, GHARCIK, School of Agriculture, University of Cape Coast, Cape Coast, Ghana.

8. Indonesian Resource Center for Indigenous Knowledge (INRIK): Prof. Dr. Kusnaka Adimihardja, Director, INRIK, Department of Anthropology, University of Padjadjaran, Bandung 40115, Indonesia.

9. Mexico Research, Teaching and Service Network on Indigenous Knowledge (Red de Investigación, Docencia y Servicio en Conocimientos Autoctonos; RIDSCA): Dr. Antonio Macias-Lopez, Director, Colegio de Postgraduados,

Indigenous Knowledge for Agricultural Development

CEICADAR, Apartado Postal L-12, C.P.72130, Col. La Libertad, Puebla, Pue., Mexico.

10. Philippine Resource Center for Sustainable Development and Indigenous Knowledge (PhiRCSDIK): Dr. Rogelio C. Serrano, National Coordinator, Philippine Council for Agriculture, Forestry, and Natural Resources Research and Development (PCAARD), Los Banos, Laguna, Philippines.

11. Kenya Resource Center for Indigenous Knowledge (KENRIK): Dr. Mohamed Isahakia, Acting Director, The National Museums of Kenya, P.O. Box 40658, Nairobi, Kenya (e-mail kenrik@tt.gn.apc.org or kenrik@tt.sasa.unep.no).

12. Sri Lanka Resource Center for Indigenous Knowledge (SLARCIK): Dr. Rohana Ulluwishewa, Director, Department of Geography, University of Sri Jayawardenepura, Nugegoda, Sri Lanka (e-mail rohana@sjp.ac.lk).

13. Venezuelan Secretariat for Indigenous Knowledge and Sustainable Development (VERSIK): Dr. Consuelo Quiroz, Coordinator, Center for Tropical Alternative Agriculture and Sustainable Development, Agrarian Science Department, Nucleo "Rafael Rangel," Universidad de Los Andes, Trujillo-Estado Trujillo, Venezuela (e-mail cquiroz@ing.ula.ve).

14. Burkina Faso Resource Center for Indigenous Knowledge (Center Burkinabè de Recherche sur les Pratiques et Savoirs Paysans; BURCIK): Dr. Basga E. Dialla, Director, IRSSH, B.P. 5154, Ouagadougou, Burkina Faso.

15. South African Resource Center for Indigenous Knowledge (SARCIK): The Institute for Indigenous Theory and Practice, 45 Castle St. (6th floor), 8001 Cape Town, South Africa (e-mail hansn@iaccess.za).

16. Brasil Resource Center for Indigenous Knowledge (BRARCIK): Dr. Antonio João Cancian, Director, Depto. de Biologia-UNESP, 14870.000, Jaboticabal-SP, Brasil (e-mail uejab@brfapesp.bitnet).

17. Nigerian Resource Center for Indigenous Knowledge (NIRCIK): Dr. James O. Olukosi, Coordinator, Institute for Agricultural Research, Ahmadu Bello University, PMB 1044, Zaria, Nigeria.

18. Uruguay Resource Center for Indigenous Knowledge (URCIK): Mr. Pedro de Hegedus, Coordinator, Centro de Estudios para el Desarrollo-Uruguay (Center for Development Studies–Uruguay; CEDESUR), Casilla Correo 20.201—Codigo Postal 12.900, Sayago, Montevideo, Uruguay (e-mail dgsa@chasque.apc.org).

19. Cameroon Indigenous Knowledge Organization (CIKO): Professor C. N. Ngwasiri, Director, Private Sector Research Institution, P.O. Box 170, Buea, Southwest Province, Cameroon.

20. Madagascar Resource Center for Indigenous Knowledge (MARCIK):

Ms. Juliette Ratsimandrava, Director, Centre d'Information et de Documentation Scientifique et Technique, Ministère de la Recherche Appliquée au Développement, 21 rue Fernand Kasanga, B.P. 6224, Antananarivo 101, Madagascar.

21. Masailand Resource Center for Indigenous Knowledge (MARECIK): Dr. Nathan Ole Lengisugi, Director, Simanjiro Animal Husbandry Vocational Training Center, P.O. Box 3084, Arusha, Tanzania.

22. Center for Indigenous Knowledge on Population, Resource and Environmental Management (CIKPREM): Professor D. S. Obikeze, Director, Department of Sociology and Anthropology, University of Nigeria, Nsukka, Nigeria.

23. Interinstitutional Consortium for Indigenous Knowledge (ICIK): Dr. Ladi Semali, Director, Pennsylvania State University, 241 Chambers Building, University Park, PA 16802, U.S.A. (e-mail LMS11@psuvm.psu.edu).

24. Georgia Resource Center for Indigenous Knowledge (GERCIK); Dr. David Kirvalidze, Director, Institute of Botany, Georgian Academy of Sciences, 22 Mosashvili St., 380062 Tbilisi, Georgia (e-mail dato@botany.kheta.ge).

25. Center for Advanced Research of Indigenous Knowledge Systems (CARIKS): Dr. Jan Brouwer, Director, P.O. Box 1, Saraswathipurm, Mysore 570 009, India.

26. Russian Resource Center for Indigenous Knowledge (RURCIK): Dr. Yevgeny Fetisov, Director, EkoNiv, P.O. Box 1 , Nemchinovka-1, Moscow Region, Russia 143013 (e-mail 100630.157@compuserve.com).

27. Center for Cosmovisions and Indigenous Knowledge–Northern Ghana (CECIK): Dr. David Millar, Director, c/o T.A.A.P., P.O. Box 42, Tamale, N/R, Ghana (e-mail aispcg@ncs.com.gh, attn. D. Millar).

28. Center for Indigenous Knowledge on Indian Bioresources (CIKIB): Dr. S. K. Jain, Director, Institute of Ethnobiology, c/o National Botanical Research Institute, P.O. Box 436, Lucknow 226001, India (e-mail manager@nbri.sirnetd.emet.in).

29. Elliniko Resource Center for Indigenous Knowledge (ELLRIK): Dr. Christos Lionis, Director, Department of Social Medicine, University of Crete, P.O. Box 1393, Heraklion, Crete, Greece (e-mail kionis@/fortezza/uer.gr).

30. Center for Indigenous Knowledge at Fourah Bay College (CIKFAB): Dr. Dominic T. Ashley, Director, Department of Sociology, Fourah Bay College, University of Sierra Leone, Private Mail Bag, Freetown, Sierra Leone.

31. Indigenous Resources Study Center (INRESC): Dr. Tesema Ta'a, Director, College of Social Sciences, Addis Ababa University, P.O. Box 1176, Addis Ababa, Ethiopia.

32. Center for Indigenous Knowledge in Farm and Infrastructure Man-

agement (CIKFIM): Dr. G. B. Ayoola, Director, Center for Food and Agricultural Strategy, University of Agriculture Makurdi, Nigeria, Private Mail Bag 2373, Makurdi, Nigeria.

IK Resource Centers Being Established

1. Regional/Subregional Centers: South Pacific Island States, Southern Africa.

2. National Centers: Benin, Namibia, Zimbabwe, Costa Rica, Colombia, Peru, Bolivia, Nepal, Australia, Vietnam, Mali, Senegal, Pakistan, Thailand, Papua New Guinea, Sudan, Ukraine.

8

Local Knowledge Systems in Latin America

CURRENT TRENDS AND CONTRIBUTIONS
TOWARD SUSTAINABLE DEVELOPMENT

Consuelo Quiroz

Landscapes of Latin America have been used and, in many places, continue to be used intensively for agriculture and agroforestry by indigenous peoples and local farmers. These people utilize sustainable systems of agriculture and agroforestry that allow them to survive on marginal farming lands (Erickson 1994). A major feature of traditional agroecosystems is their high degree of genetic and cultural diversity (Altieri 1987). The Andes and Amazonia, for example, are two zones with the highest biological and cultural diversity in the world (Altieri, Anderson, and Merrick 1987). People inhabiting these areas have developed traditional production systems that seek a dynamic equilibrium between society and nature and that provide good examples of organizational,

This chapter was written during a sabbatical year 1994–1995. Thanks are gratefully offered to the University of the Andes and the Kellogg Foundation for their financial support during that period. I am also grateful to Dr. Michael Warren and his staff at the Center for Indigenous Knowledge for Agriculture and Rural Development (CIKARD), Iowa State University, for their kind and valuable support.

social, and technological creativity (Altieri, Anderson, and Merrick 1987; Rist 1991b; Warren 1992).

At present, these agricultural systems are rapidly disappearing, as is the important knowledge on sustainable natural resource management associated with them (Cunningham 1991; Erickson 1994; Haverkort and Millar 1994). One of the major reasons for this rapid loss is the poverty of vast sectors of the population and the concomitant increasing pressure on natural resources. Furthermore, the introduction of "modern" market-oriented agricultural and forestry technology is not only displacing, even eliminating, local practices in favor of high input agriculture (monocropping, large-scale commercial farming, and cattle raising), it has also brought with it the degradation of natural resources, poorer nutrition, and the loss of informal channels of communication (Gliessman, Garcia, and Amador 1981; Shiva 1993).

Sustainable development approaches have recently been proposed as a potential answer to the problems of natural resource deterioration and mass poverty. As a consequence, there has been an upsurge in the study of local knowledge systems by research scientists and extension workers in developing countries.[1] These researchers recognize the importance of understanding, respecting, and utilizing holistic views of natural resource management such as those found in local agroecological knowledge systems. The agroecosystems based on this knowledge are proving significant because of their ability to supply high outputs of calories and proteins without damaging the resource base and for assuring, in general, sustainability of development (Brokensha, Warren, and Werner 1980; Hecht and Posey 1989).

This chapter presents an overview of the literature on current trends among projects that have documented or utilized local knowledge systems (LKS) for development purposes in Latin America.[2] It also will look at some specific examples of LKS, exploring their contributions—both real and potential—to sustainable development approaches in the region. Discussion will then focus on the issue of gender and how it relates to local knowledge systems and natural resource management. Finally, some conclusions will be presented, with recommendations and policy implications.

Current Trends

There are now hundreds of studies of projects that have documented local knowledge systems around the world, including Latin America. One way of classifying these studies is to take into consideration the extent to which the people studied ("beneficiaries") have been involved in the design and implementation stages of the projects. Three broad study types can be considered: (1) grassroots projects, with participation high; (2) rural development and research projects, with participation medium to high; and (3) research projects, with participation low to none.

Grassroots Projects

These are development efforts designed and implemented by the beneficiaries themselves—sometimes with motivation and help from outsiders. An example of a grassroots project is the reintroduction of indigenous foods by a group of women in Independencia (Ayopaya Province, Department of Cochabamba, Bolivia) who worked together to create a small enterprise for processing Andean cereals (mainly quinoa) into noodles. Quinoa is a nutritious cereal, first domesticated in the Andes as early as five thousand years ago. It has long been a staple food in the Bolivian highlands, but colonial influences and modernization have influenced its patterns of utilization. In 1991 the women of Independencia received financial support from a German agency that enabled them to buy a house and an Italian noodle machine for their enterprise (Villavicencio, Aquino, and Aquino 1993).

Another example of grassroots development is a project focusing on the commercialization of nontimber forest products extracted by a group of indigenous women from communities in the Brazilian state of Acre. In 1989 with the assistance of the Rural Women's Group, affiliated with the Rural Workers Union of Xapuri, the women organized two product fairs, where they sold products they had collected from the forest and processed. These included jams, basketry, rubber goods, juices, nuts, and honey. The economic importance of processing is considerable, as processed goods carry a much higher market value than unprocessed goods. For example, at the product fairs, one kilogram of raw Brazil nuts was priced at $0.36—compared to $3.60 for one kilogram of chocolate-covered Brazil nuts (Kainer and Duryea 1992).[3]

Rural Development and Research Projects

These are usually small-scale projects whose main purpose is to improve the living conditions of farmers and other rural peoples through the implementation of participatory research and extension approaches. These projects to some extent incorporate elements of local knowledge systems in their work. Some of them have tried to probe traditional practices scientifically, and others have included components of local knowledge systems in the curricula of educational programs. In this category, we find projects that have been carried out by national universities (either alone or in collaboration with foreign universities), by international research centers, and by other institutions.

For projects undertaken by national universities, the Agroecological Project of the University of Cochabamba (AGRUCO), Bolivia, initiated in 1985, serves as an example. Its aim has been "to contribute to the learning process at universities and other institutions involved in rural development" (Rist 1991b:22). To achieve this aim, the project has worked in several Bolivian Andean communities as a small participatory research and extension program. Some indigenous techniques that have been studied scientifically through this project include crop rotation, the use of organic fertilizer (manure), mixed cropping, minimum tillage in dry areas, and the use of weather-forecasting indicators before plowing (Rist 1991a, 1991b, 1994).

Another example at this level is the Foundation for the Application and Teaching of the Sciences (FUNDAEC), created in Cali, Colombia, in 1974 as an interdisciplinary group of the Universidad del Valle. FUNDAEC has been working in the state of Cauca with the goal of building "a harmonious combination of external and local knowledge to create more efficient production systems that accord with the farmers' decision-making" (Torné de Valcárel 1991:15). Through one of their programs, called the Rural University, they have designed, implemented, and disseminated in different regions of the country a three-level secondary school program (the Tutorial Learning System) on "rural well-being," which is adapted to the general and specific conditions (geographic and sociocultural) of each zone. The system is designed so that young campesinos do not have to abandon their productive activities. Recently, a university degree in Rural Education was created, to prepare individuals to further disseminate the advantages of these learning processes (Torné de Valcárel 1991).

An example of a collaborative approach among universities is the Small Ruminant Collaborative Research Support Program (SRCRSP), an interdisciplinary, interinstitutional research and development project carried out during the 1980s in Peru by the Universidad de San Marcos, Universidad Nacional Agraria de la Molina, the University of Missouri, Columbia University, Winrock International Institute, and several NGOs such as Grupo Yanapai. In this participatory program, numerous traditional peasant practices—primarily involving ethnoveterinary medicine, livestock and crop management, social organization, and gender—were recorded and studied, with some being scientifically tested (Fernández 1989; Jiménez and Hobbs 1985; Mathias-Mundy 1989; Mathias-Mundy and McCorkle 1989; McCorkle 1991).

For projects undertaken by international research centers, an example is the work carried out by the International Center for Tropical Agriculture (CIAT) in Colombia. CIAT has undertaken a series of studies that systematically involve small farmers as participants in the planning, execution, and evaluation of research. The work has led to the generation of specific methodologies for participatory technology development (Ashby, Quirós, and Rivera 1987; Ashby and Pacheco 1988). Another example of this approach is the work done by the International Potato Center (CIP) in Peru, where, through the evaluation of projects that have utilized Farming System Research and Extension (FSRE) approaches (for example, Mantaro Valley), the importance of including farmers' knowledge and native potato varieties through the different stages of projects has become clear (Horton 1986).

For projects undertaken by other types of institutions, the approaches employed are illustrated by the work of the Brazilian Institute of Environment and Natural Renewable Resources (IBAMA), established in 1984. In 1992 IBAMA started a project with support from the Food and Agriculture Organization (FAO) to introduce "environmental education" into the school curriculum in Rio Grande do Norte, a poor state in northeastern Brazil. This integrated program involves the teachers and their supervising officers as well as agricultural extension officers. One characteristic of this project is "its ability to base its activities on the reality in the community and create a forum where community members are also recognized as important partners in the efforts to improve the environmental situation" (Grit 1993:10).

Research Projects

These are generally small scale and have as their main objective the recording, documenting, and study of local knowledge systems. These studies are not necessarily linked directly to a particular development project. Under this category, we find researchers (individually or in teams) working in universities, national indigenous centers, and national resource centers for indigenous knowledge. The majority of documented studies of LKS in Latin America fall within this category. These studies have documented many different kinds of LKS (such as ethnobotanical, ethnoecological, ethnoveterinary, ethnoentomological, and ethnomedicinal) possessed and used by groups of indigenous people and farmers inhabiting the diverse landscapes of Latin America. A few examples are studies focusing on the Kayapó Indians of Brazil (IPGRI 1994; Hecht and Posey 1989; Posey 1982, 1983, 1985, 1991; Posey et al. 1984); the Machiguenga of Peru (Johnson 1983); the Totonacas of Mexico (Medellín-Morales 1990); the Yarimaguas of Peru (Bidegaray and Rhoades n.d.); the Huastec of Mexico (Alcorn 1984); the Aynokas of Bolivia (Rist 1994); Andean potato growers (Benzing 1989; Brush 1980); the Bora of Peru (Alcorn 1984); the Yukuna of Colombia (Reichel 1992); indigenous women of the Chiapas highlands in Mexico (Pérezgrovas, Pedraza, and Peralta 1992); experimentation by farmers in general (Quiroz 1995; Rhoades and Bebbington 1995; Warren, Fujisaka, and Prain 1995); and maize farmers in Mexico (Altieri and Trujillo 1987; Bellon and Brush 1994).

LKS Contributions Toward Sustainable Development

This section provides an overview of some of the LKS that have been studied in Latin America. It is proposed that these can potentially contribute to enhancing sustainable development and poverty alleviation approaches in the region. Examples of local knowledge systems re-viewed here fall under several categories: (1) in situ biodiversity con-servation by traditional farmers; (2) raised-field technology; (3) land resource management by indigenous people in Amazonia; (4) eth-noveterinary medicine.

In Situ Biodiversity Conservation by Traditional Farmers

The peasant mode of production, when not disrupted by external eco-nomic or political forces, is generally characterized as having a high degree of diversity and as using production systems (such as terraces and hedgerows in sloping areas, minimal tillage, small field sizes, and long fallow cycles) that cause minimal land degradation (Altieri, An-derson, and Merrick 1987; Chambers, Pacey, and Thrupp 1989). Farm-ers can act both directly (through seed selection and the maintenance of specific varieties for different soil conditions) and indirectly (through planting patterns) on the genetic variability of their fields (Bellon and Brush 1994). These diverse agroecosystems are usually composed of cultivated and fallow fields, complex home gardens, and agroforestry plots that generally contain more than one hundred plant species per field. The plants have multiple uses—as construction materials, for ex-ample, or as firewood, tools, medicines, livestock feed, and human food. Such genetic diversity allows farmers to mitigate, at least partially, against plant diseases, while exploiting numerous microclimates (Altieri, Anderson, and Merrick 1987; Chang 1977; Clawson 1985). Many of these traditional agroecosystems are still found throughout the land-scapes of developing countries and can constitute major in situ reposito-ries of both domesticated and nondomesticated plant germplasm (Altieri 1987; Altieri, Anderson, and Merrick 1987; Bellon and Brush 1994).

Several studies on the diversity of peasant agroecosystems in Latin America have focused on maize in Mesoamerica and potatoes in the Andean region, because these regions have long traditions in growing these crops. One example is the study by Bellon and Brush (1994) of maize farmers in the central Chiapas region of southern Mexico. The authors found that the farmers (in spite of the widespread adoption of a modern, high-yielding variety of maize) continue to select local varieties for specific soil types. These maize growers keep local varieties (landraces) belonging to six races and four race mixtures. Fifteen local varieties are recognized (no one variety alone satisfies all of the farmers' concerns). Another example is from the work of Medellín-Morales (1990) on agrosilvicultural management by Totonaca Indians in Mexico. These indigenous people generally manage and use an average of 234 plant species, 110 animal species, and 39 fungus species. They have a profound knowledge of soils (they recognize twelve different types) and are aware of the successional changes of vegetation and weather through the year.

In relation to potato growers, the study by Benzing (1989) in the village of Boliche in the Ecuadorian highlands serves as an example. Benzing found that these farmers grow an average of twenty different potato varieties in the same plot and are very knowledgeable about each variety's resistance and susceptibility to diseases and climatic influences. They grow the potatoes in rotation with three other Andean tubers.

Many of these traditional agroecosystems are at risk because of economic and political forces that push farmers to accept newly introduced varieties and "modern" technologies. Research indicates that this trend has not only resulted in the disappearance of indigenous crop varieties and potentially useful germplasm, but, as peasants abandon subsistence crops to produce cash crops or are deprived of sufficient land and forced to work as wage laborers, the resulting economic, social, ecological, and dietary changes often lead to poorer health and nutrition (Altieri and Merrick 1987; Chambers, Pacey, and Thrupp 1989). One way of dealing with this complex problem (loss of diversity and impoverishment of small farmers) would be to combine preservation efforts with rural development projects that take into account the ethnobotanical knowledge of rural people and that emphasize both food self-sufficiency and local resource conservation (Altieri and Merrick 1987; Altieri, Anderson, and Merrick 1987).

There are lessons to be learned from the limited number of grassroots rural development programs currently operating in the Third World. Altieri and Merrick (1987:93) suggest:

the process of agricultural betterment must (a) utilize and promote autochthonous knowledge and resource-efficient technologies, (b) emphasize use of local and indigenous resources, including valuable crop germplasm as well as essentials like firewood resources and medicinal plants, and (c) be a self-contained, village-based effort with the active participation of peasants. The subsidizing of peasant agricultural systems with external resources (pesticides, fertilizers, irrigation water) can bring high levels of productivity through dominance of the production system, but these systems are sustainable only at high external cost and depend on the uninterrupted availability of commercial inputs. An agricultural strategy based on a diversity of plants and cropping systems can bring moderate to high levels of productivity through manipulation and exploitation of the resources internal to the farm and can be sustainable at a much lower cost and for a longer period of time.

Raised-Field Technology

The management of regularly inundated lowlands, salinity-affected areas, or tidal swamps for agriculture by using raised beds or fields has been practiced extensively by indigenous peoples in Latin America and in other tropical regions throughout the world for at least two thousand years. The practice seems ancient. Raised bed farming was developed in China in the fifth century B.C. (Reijntjes 1991; Thurston and Parker 1995).

Erickson (1994:3) defines raised fields as "large planting surfaces of earth elevated above the seasonally flooded savannas and wetlands." The basis for these methods is balanced water management through systems of ridges and furrows. The furrow soils are used to create well-drained raised fields for the cultivation of a wide variety of crops and trees. The furrows function as drainage canals; in some places they are used also for the cultivation of wetland rice and for aquaculture. Sediments from floods, organic matter from aquatic plants, soil moisture-nutrient dynamics between ridges and furrows, favorable microclimates induced by water and trees, and reduced risk of complete crop failure, all contribute to the sustainable success of these relatively highly productive systems (Reijntjes 1991). In the *waru waru* system of Peru, the raised beds contribute to frost control as well as to better water management. Some studies have shown that, in raised bed systems such as the Mexican *chinampas*, elevated biological activity results in less incidence of diseases (Thurston and Parker 1995).

Until recently, prehispanic raised fields were relatively understudied. Some authors (Erickson 1994) have described the approximately 170,000 hectares of raised-field remnants found in South America. Extensive systems of intensive raised-field management were found in Mexico (where they are known as chinampas) and also were common in Central America. North American Indians used raised-bed agriculture in several regions before the arrival of Europeans (Erickson 1994). Chinampas were probably first developed by the Maya and subsequently used by other Indian cultures in Mexico and Central America. A system similar to chinampas was being used by Venezuelan Indians on the swampy shores of Lake Maracaibo when the Spanish arrived (Thurston and Parker 1995). Various authors have described over eighty thousand hectares of raised fields called *waru waru* or *camellones* at 3,660 meters above sea level near Lake Titicaca on the border between Peru and Bolivia (Thurston and Parker 1995).

Studies have found that some of these systems sustained large populations organized in sizable villages and towns dispersed throughout savannas and forests. Several thousand "islands"—low mounds, several hectares in surface area—are found in the savannas. Raised fields provided a stable and (presumably) sustainable production base for ancient inhabitants of the region (Erickson 1994). The high productivity of chinampas has been cited as a major factor in allowing the Aztecs to grow from a small tribe to a powerful ethnic group that essentially dominated most of Mexico when the Spanish arrived. Some authors have estimated that the Aztec chinampas may have been capable of feeding one hundred thousand people (Thurston and Parker 1995).

In experiments where raised fields have been rebuilt according to specifications obtained from archeological studies, higher crop yields have been obtained from the raised fields than from nonraised fields in neighboring areas—for example, potato yields of eight tons per hectare for the raised fields compared to two or three tons per hectare average from nonraised fields in the Puno district of Peru (Thurston and

Parker 1995). Several attempts have been made to transplant raised-field technology to other locales, with less positive results. In Mexico, the raised fields failed because the model was transplanted without considering the specific ecological, social, economic, and political contexts of the raised-field agroecosystems. In Tabasco, for example, two projects carried out by the National Institute of Biotic Resources (INIREB) in the 1970s among the Chontal Indians were designed and implemented by outside technicians, without significant local participation. In general, the technicians overlooked the wider socioeconomic and political context in which the farmers lived. They ignored the fact that labor was scarce and its cost very high because of the discovery of oil reserves in the area, and they did not pay attention to the lack of marketing opportunities for produce (Chapin 1988; Reijntjes 1991; Thurston and Parker 1995).

Another attempt to transplant the raised-field model was made in 1990 by the Agro-Archeological Project of the Beni in Bolivia—a collaborative research project between the Bolivian National Institute of Archeology and the University of Pennsylvania (Erickson 1994). In this project, a small block of raised fields (0.5 hectares) was constructed in the pampa at the biological station of the Beni. The crops (maize and manioc) grown on the raised fields had higher yields than those grown in the surrounding area. The project technicians also attempted to convince farmers to build raised fields through a highly participatory (bottom-up) approach. This part of the project was carried out in collaboration with a small (two-person staff) grassroots development organization that had recently received a grant from the Inter-American Foundation for promoting the adoption of raised fields. Incentives offered to the farmers to encourage their participation included a small daily wage (approximately two dollars a day), seeds, the loan of tools, and food for lunch during workdays. Moreover, participants received all crops produced on the fields. In return, the farmers would provide land and labor and would help monitor the fields and collect data. The response of the community was very positive, although the scarcity of labor was a serious problem in the area. The community fields were planted with local crops, including maize, manioc, beans, peanuts, bananas, sweet potatoes, and wet rice (the latter being an innovation suggested by the community members that would yield an additional harvest) (Erickson 1994). Another attempt, sponsored by the Peruvian government, involves a program that gives farmers credit to reconstruct raised fields—approximately fifteen hundred hectares have already been reconstructed through this project (Thurston and Parker 1995).

Sufficient data on long-term experiments with raised-field agriculture are not yet available to indicate whether and under what conditions it might prove truly sustainable (Erickson 1994). Opinions are divided about the future of raised-field agriculture in Latin America. Some authors, like Chapin (1988:16), believe that although the chinampas system has survived into the twentieth century, "it has done so only as a threadbare remnant [and] . . . it will soon exist only in history books." According to Chapin, chinampas were viable in the prehispanic era because "the total ecological context—including the dynamics of human society as well as that of 'nature'—was right."

On the other hand, scholars such as Erickson (1994) believe that raised-field technology represents a potentially sustainable solution to the environmental and technical problems facing agricultural production. Citing the work of cultural geographer William Denevan (1966:4–5), Erickson (1994:10) indicates that raised-field technology has significant promise for utilization in seasonally inundated savanna, river floodplain, and permanent wetland zones in the humid tropical lowlands of South America, including the llanos region of the Orinoco River in Venezuela, the Pantanal of western Mato Grosso in Brazil, eastern Marajó Island in Brazil, the Bolívar savannas of northern Colombia, the Amazon floodplain, the Casiquiare River region in Brazil and Venezuela, the Heath River in southeastern Peru, interior and coastal savannas of Guiana, the Gran Chaco of Paraguay and Bolivia, and parts of the Planalto Central in Brazil.

Natural Resource Management by Indigenous People in Amazonia

Amazonia constitutes the largest area of tropical forest in the world—over fifty-five million hectares (Posey et al. 1984). These tropical forest ecosystems have been shown to be very fragile and prone to collapse if disturbed. During the last twenty-five or thirty years, more than twenty million hectares of lowland forests in the Amazon basin have been transformed through human activity to other land uses. For instance, in the Colombian Amazon more than six million hectares have been converted into poor quality pastures for cattle ranching (Bunyard 1994). The main forms of regional agriculture that follow forest conversion are very unstable and prone to produce deterioration of soil fertility, which in turn causes agricultural failure. As soil nutrients are exhausted and the costs of weeding become high, farmers and ranchers abandon old areas and clear new ones, thus creating an ever expanding frontier of forest removal and land degradation (Hecht and Posey 1989). In general, so-called development projects in the region have resulted in vast deforestation, soil destruction, and desertification (Posey et al. 1984).

Parallel to these environmental disasters are the drastic effects this "development" process has had on the indigenous populations who have inhabited the area for millennia. Entire ethnic groups have been exterminated, and those remaining risk imminent demise (Bunyard 1994; Posey 1982). For example, Posey et al. (1984) reported that over the last seventy-five years in the Brazilian Amazon alone, at least eighty-seven Amerindian societies have become extinct. With the extinction of each indigenous group, the world loses millennia of accumulated knowledge about life in and adaptation to tropical ecosystems (Posey 1991).

Recent studies indicate that the traditional view of the Amazonian environment as a static entity is incorrect (Erickson 1994). For thousands of years, Amerindians have survived in Amazonia thanks to their understanding of ecological zones, plant-human-animal relationships, and NRM strategies that evolved through complex processes of trial-and-error experimentation (Posey 1985, 1991). Undoubtedly, the local knowledge possessed by these indigenous peoples offers a rich, untapped source of information about how to manage, conserve, and utilize—in a sustainable manner—the natural resources of the Amazon Basin (Posey et al. 1984).

The local NRM knowledge systems of some Amazonian groups have been studied in detail. The work done by anthropologist Darrell Posey among the Northern Kayapó in Brazil since 1977 is a case in point. The Gorotire Kayapó have inhabited their present territory for centuries. In pre-Columbian times, they were a large group (exceeding ten thousand) even though they supported themselves on semidomesticates for one-half to three-quarters of each year. By the mid–twentieth century, they had been reduced to a population of about five hundred. In 1952 the Brazilian government established a two-million-hectare reserve in the Brazilian state of Pará in the hope of ensuring their survival. There are currently nine villages scattered over the reserve, with a population of approximately twenty-five hundred (Posey 1982, 1991).

The knowledge of the Kayapó about natural resource management, conservation, and utilization represents an integrated system of beliefs and practices. The specialized burning and planting strategies they employ are sophisticated adaptations to the tropical ecological zones of the Xingu River basin (Posey 1982:24). In addition to information shared generally among the Kayapó, there also is specialized knowledge held by only a few individuals. Each village has its specialists in soils, plants, animals, crops, medicines, and rituals. The Kayapó's ethnoecological knowledge is profound—they recognize different kinds of ecosystems that fall between the poles of forest and savanna. Their agriculture focuses on the zones intermediate between forest and savanna, because it is there that maximal biological diversity occurs. They not only recognize the richness of these zones, they actually create some of these zones through management techniques. They exploit secondary forest areas and create special concentrations of plants in forest fields, rock outcroppings, trail sides, and elsewhere. Research on the Kayapó shows that they base forest management on a view of plant communities as integrated systems rather than as individual species. Moreover, manipulated wildlife and even semidomesticated bees each play a role in their overall management strategies (Posey 1985, 1991).

It is important to emphasize that studies also have shown that not all native Amazonians are conservationists by nature—particularly in light of changes occurring in the region (see Stearman 1994 on the Yuquí of Bolivia). Resource management systems, like those of the Kayapó, come from areas where human survival depends on the careful management of the subsistence base. Stearman (1994:2) notes, "it should not be taken as a 'conservation ethic' possessed by all the Indians who inhabit Amazonia, but it is the requirement for resource management through specific adaptive strategies."

The Kayapó are only one of many indigenous groups inhabiting Amazonia, but, Posey (1985:156) argues, "the lessons they have learned through millennia of accumulated experience and survival are invaluable to a modern world in much need of rediscovering its ecological and humanistic roots." As many of these groups and their knowledge disappear, such lessons are lost forever. Long-term sustainable management strategies, like those of the Kayapó, offer many fundamental principles that could lead to a more socially and ecologically sound development approach throughout the humid tropics (Posey 1985, 1991).

Ethnoveterinary Medicine

Ethnoveterinary medicine "deals with folk beliefs, knowledge, skills, methods and practices pertaining to the health care of animals" (Mathias-Mundy and McCorkle 1989:3). Over centuries, traditional peoples and farmers have developed a rich empirical knowledge about different kinds of cattle—including management aspects such as identification, housing, handling, feeding, health, reproduction, and disease prevention. This knowledge is based on the close observation of animals and the oral transmission of experiences from one generation to the next. It has proved generally effective—results from the laboratory have often only confirmed what many indigenous peoples and local farmers already knew from firsthand experience (Mathias-Mundy and McCorkle 1989; Pérezgrovas, Pedraza, and Peralta 1992).

An example wherein villagers helped project scientists (in this case from the SRCRSP) in problem solving comes from the use of a wild tobacco plant to cure skin disease. The scientists helped farmers to test the efficacy of a tobacco dip in reducing the number of parasites infecting animals' skins. They found that the reduction was greater than 90 percent. Another example stems from the traditional practice of using artichoke leaves as cattle feed to prevent the animals from getting liver flukes. Experiments revealed that the leaves did indeed reduce the incidence of these debilitating parasites (Mathias-Mundy 1989).

Traditional veterinary practices offer a particularly rich resource for development efforts. Where livestock owners live too far from a clinic or cannot afford to pay for veterinarian fees and medication, they are forced to rely on traditional treatments. It is, of course, important to emphasize that traditional veterinary practices are not a panacea. There are limitations to the efficacy of ethnoveterinary medicine: the inconvenience of preparing or using some remedies, the seasonality of certain medicinal plants, the lack of treatments against infectious epidemic diseases, the ineffectiveness or even harm caused by some treatments, and inadequate diagnosis (Mathias-Mundy 1989; Mathias-Mundy and McCorkle 1989).

LKS, NRM, and Gender

Many women in rural areas of Latin America (as in many other regions of the world) perform roles that are closely related to the conservation and utilization of natural resources. For example, in Colombia, land preparation, planting, weeding, and harvesting of small plots is done primarily by women. In most Latin American countries, home gardens and small animals (chicken, goats, pigs) are cared for by women and children. In some countries, it is common to find women processing crops—in Cuba, for example, women grind corn and string tobacco to dry. Throughout South America, the processing of bitter manioc is almost universally done by women. In some Latin American societies, women actually devote more time to agricultural activities than do men. In Colombia, one study showed that men spent 5.78 hours per day on "productive work," whereas women worked 6.56 hours per day (Fortmann 1989:39). Nonetheless, although women play such important roles in natural resource management, the gendered nature of local knowledge systems in the management of resources and the maintenance of biodiversity remains generally unrecognized or underestimated. As a consequence, women's needs and the knowledge they possess have usually not been issues in mainstream sustainable development circles (Appleton et al. 1993; Awa 1989; Badri and Badri 1994; Jiggins 1994; Norem, Yoder, and Martin 1989; Quiroz 1994; Rocheleau 1991; Shiva and Dankelman 1992).

Since knowledge is socially constructed and because gender is one of the primary dimensions of social relations, it is easy to see why knowledge is gender-based (see Norem, Yoder, and Martin 1989). Because of the division of socioeconomic responsibilities (which should not be viewed as static) between men and women, some types of knowledge will be primarily the domain of women, others primarily the domain of men, and still others will be complementary. In some of these domains, women and men will have the same interests and needs, but in others there may be conflicts. Social differences (including age, origins, descent position, or socioeconomic status) are essential conditioners of the way natural resources are managed. These differences interact with gender to shape people's resource management strategies and to control claims at the farm, community, and landscape levels (Jiggins 1994; Leach 1994; Simpson 1994).

The knowledge possessed and utilized by women is closely related to their daily activities. These experiences and the way women interact with others and with the natural environment give them a different perspective, a different kind of knowledge—what Jiggins (1994) calls "distinctive knowledge." A major feature of distinctive knowledge is that it tends to be more holistic, especially in comparison to knowledge possessed by men involved in market-oriented agriculture (Jiggins 1994; Shiva 1993).

Several studies have focused on women's roles and their knowledge related to natural resource management in Latin America. Research on the livestock-related knowledge possessed by indigenous women in the highlands of Peru was undertaken by the SRCRSP. This research showed that women and children are the ones who take care of animals (mainly sheep, pigs, cows, and donkeys), while crops (mainly potatoes, barley, wheat, oats, and alfalfa) are primarily the domain of men (Fernández 1989).

A study on the ethnobotanical knowledge of indigenous women from reserves in Acre, Brazil, was carried out by Kainer and Duryea (1992). They found that these women possess considerable skills and knowledge regarding the plant resources of the reserves, as well as a refined ability to identify, cultivate, manage, and process a large number (over 150 wild and domesticated species) of "minor" plants. The

Local Knowledge Systems in Latin America

women exhibit particular proficiency in plant processing, especially of species used for food, spices, beverages, and medicines.

Fortmann and Rocheleau (1985) investigated the involvement of women in agroforestry projects (Plan Sierra) in the Dominican Republic and found that, during the first three years of the project, women were included in only a limited number of program activities. However, their involvement in other, "invisible," activities was overlooked. The study showed that women were actually involved in a wide variety of activities, including harvesting of annual crops and coffee (as owners or hired farm workers), raising small animals such as pigs and chickens for meat and egg production, milking cows for home consumption, and caring for home gardens (usually vegetables, bananas, and herbs). Fuelwood and water collection were primarily the responsibility of the women, with some help from children. Cheese, candy, and cassava processing were almost entirely women's enterprises. Palm fibre containers were produced and sold by the women, with help from children and the elderly. Women artisans wove the seats and backs onto locally manufactured wooden chair frames, with subcontract agreements with men's woodworking shops.

When the Institute for High Altitude and Tropical Veterinary Research (IVITA) in Huancayo, Peru, began its research on small-scale mixed farming systems in the Mantaro Valley of Peru, it was assumed that men were in charge of all on-farm decisions. Thus, most of the livestock extension, training, and development programs had men as their target group. After a number of years, however, it became clear that women participated in a wide range of on-farm activities, especially animal production. This research indicates that women possess a great deal of local knowledge about animal health, nutrition, reproduction, and selection. Women make management decisions related to the selection of animals for sale, fodder acquisition, the treatment of diseases, and the organization and control of rotational grazing strategies (Fernández-Baca 1994).

It is important to stress that a focus on gender issues in local knowledge systems does not imply focusing only on women's knowledge: "Women's relations with men condition what they are doing and why. . . . Women perform those tasks (e.g. fuel and food procurement) because of social, not 'natural' divisions of labor" (Leach 1994:213). Therefore, women should not be treated as independent individuals, removed from their complex range of interactions with men and the contexts in which they live (Jiggins 1994; Norem, Yoder, and Martin 1989).

Recommendations and Conclusions

The task of strengthening local knowledge systems and institutions related to natural resource management is a complex issue. It cannot be viewed in isolation from other elements (intellectual property rights or the national economy, for instance) in a sustainable development approach. There are many ways development agencies such as the World Bank could aid this process. Development agencies might lend greater support to the activities being carried out by the growing international network of regional and national indigenous knowledge resource centers (Warren 1991a). The range of activities of these centers is broad. Collaboration with development agencies could begin with a focus on some of the following ideas.

First, raising awareness of the value of indigenous and local knowledge systems can be achieved at a number of levels, including the community level. Mathias (1994:7) argues that an important starting point for LKS conservation and utilization is raising awareness of the value of this kind of knowledge among the people who have developed it, because, among other things, "getting people to recognize the value of their IK stimulates their self-confidence." The community has to be involved in developing a strategy of conservation and utilization of their LKS. One way to achieve this is by recording LKS success stories and sharing them with the community in the form of songs, drawings, puppets, plays, stories, dramas, videos, and other traditional or modern means of communication (Mathias 1994).

Second, an important level for targeting is the educational system. This can be done through the incorporation of elements from the local knowledge systems into the curricula of different programs (for extension agent training, for example, see Quiroz 1993). It would be constructive to develop and enhance local human resources—through training and extension programs, for example—to enable the community to take control of their own development efforts. It would be useful to strengthen local institutions by working with, not against, them in building on the diversity of institutions, channels of communication, and socioeconomic arrangements through which the community (including women) can and do press their interests in day-to-day life (Leach 1994).

Third, another positive approach would be to support programs that combine the preservation of resources and development efforts (Altieri and Merrick 1987; Altieri, Anderson, and Merrick 1987). Such programs should treat people as "active actors," that is, where the indigenous person or farmer is seen as "an active strategizer who problematizes situations, processes information and brings together the elements necessary for operating the farm" (Long and Villareal 1994:47). This type of approach requires several institutional changes that allow the development worker or planner to be more sensitive to local reality and needs (including women's). An example of the kind of institutional changes needed would be modifications in evaluation procedures that create mechanisms whereby the extension or development worker would be evaluated by qualitative, as well as the usual quantitative, criteria.

Fourth, it would be valuable to support knowledge sharing among different groups, including farmer-to-farmer, farmer-to-extensionist, farmer-to-planner, farmer-to-researcher, researcher-to-researcher, and so on. These groups can come from the same or different area, country, or region. The sharing could be done through means such as workshops, videos, field trips, written materials, and electronic media (Salas and Tillman 1990).

Fifth, it would be helpful to develop research on local knowledge systems. Topics for this research can be diverse and will depend on the need of each particular area, country, or region. For example, there is a need for research that helps "to improve field methodologies for recording indigenous knowledge and decision-making systems" (Warren 1991a:33). There also is a need for research about local knowledge systems that takes into account variations in resource access, use, and control according to social differences (including gender, age, origin), since each group will not only have specialized knowledge but also "will bring different interests to bear, and different rights and opportunities to access, use and control the resources and products concerned" (Leach 1994:37).

Evidence indicates that, very often, the implementation of "scientifically" based Western resource management in Latin America has not resulted in sustainable resource use. Consequently, the urgent need for alternative approaches is no longer questioned. NRM systems based on the knowledge and experiences of the resource users themselves have the potential to improve the management of a number of agroecosystems in Latin America. These local knowledge systems provide a reservoir of adaptations that may be of wide-ranging importance to the design of sustainable systems (Berkes and Folke 1994). Indigenous and local communities not only have extensive knowledge of their surrounding environments, they also are ultimately responsible for implementing conservation policies in their territories (UNEP 1994). Recognizing local ecological knowledge as the basis for any intervention, and accepting local people as resources rather than problems will allow development personnel to complement rather than try to replace the valuable knowledge and capacities developed through many years of trial-and-error adaptations by these peoples (Kilahama 1994). Through the convergence of the parallel knowledge and experiences of farmers, indigenous peoples, and development personnel, new opportunities for supporting sustainable natural resource management have the potential to emerge.

9

Biodiversity and Agroforestry Along the Amazon Floodplain

Nigel J. H. Smith

A Rationale for Conserving Biodiversity

The conservation and rational management of biodiversity is essential for long-term development. The preservation of significant areas of natural habitats—with their wide array of species—protects resources for the genetic improvement of agriculture, forestry, and aquaculture, among other economic activities (Schultes 1990; Smith et al. 1992; Wilson 1992). With ample gene pools to draw upon, plant breeders and farmers are able to respond more readily to the challenge of raising and sustaining crop yields. Options for domesticating more plants are increased if biodiversity is maintained (Schultes 1980). Ecologically diverse ecosystems provide a suite of environmental services important for the welfare of people living in them. In the case of the Amazon, for example, floodplain forests play a critical role in maintaining the productivity of the region's fisheries. The Amazon floodplain is poised to become the next agricultural frontier in northern South America, but unless carefully managed, large-scale deforestation will trigger a massive loss of biodiversity comparable to the devastation in Brazil's Atlantic rainforest. Rampant destruction of floodplain forests along the Amazon is destroying the commercial and subsistence fisheries that provide the bulk of animal protein in the region and is eliminating plants used by locals for sustenance, fish bait, fuelwood, and construction (Goulding, Smith, and Mahar 1996).

Forces of Destruction and the Need for Agricultural Intensification

The expansion of cattle and water buffalo ranching is the main cause of deforestation on the Amazon floodplain. Market forces are driving the increase in livestock herds, so no quick fixes are available, such as the removal of fiscal incentives. Along the middle Amazon, even small farmers are acquiring cattle as an investment and as a source of manure for vegetable gardens. The collapse of jute prices in the early 1980s forced farmers to seek other sources of income, and many have increased their fishing efforts or have turned to livestock production (McGrath et al. 1993). Economically viable alternatives are needed to stem the tide of destruction. Agroforestry, a much neglected option for agricultural intensification on the Amazon floodplain, could not only boost incomes of farmers, and thereby reduce pressure on the remaining forests, but could easily be integrated with fisheries and aquaculture (Smith 1999).

In order to develop effectively the agricultural potential of the Amazon floodplain, scientists will need to tap the rich gene pools of wild and cultivated plants. By harnessing this latent potential, it should be possible to develop more nutritious and productive crops, thereby providing a viable alternative to cattle raising. Farmers on the Amazon floodplain have already selected a diverse assortment of perennial crops in their home gardens, some of which could be multiplied for more widespread planting or could be used by crop breeders to produce more disease- and pest-resistant varieties.

Survival of floodplain forests is essential for several reasons. First, floodplain forests provide a wide array of plant products for small farmers and fishermen living along the Amazon. The disappearance of forests will force many people with already modest incomes to purchase some of the lost products in markets. Second, the fate of forests and the future of agroforestry are linked. Many wild plants in the forest are being domesticated in home gardens, thus providing the next generation of crops for inhabitants of the floodplain and elsewhere. Third, wild populations of some existing crops are found in floodplain forests, thus rampant deforestation could extinguish useful genes for crop improvement (Smith and Schultes 1990).

This chapter documents some of the uses of floodplain forests by locals, explores the importance of home gardens, and assesses the potential for large-scale agroforestry for multiple products. The key to successful development of natural resources on the Amazon floodplain is the blending of traditional knowledge with modern science (Pichón and Uquillas 1995). A working assumption here is that some resource management strategies that boost incomes while sparing the environment have already been worked out by farmers on the floodplain, and that, if better understood, several of these promising practices could be adapted for much wider dissemination. For example, a cornucopia of fruit and nut species are grown on a small scale in home gardens. However, many of these species have the potential to be cultivated in plantations. Preliminary fieldwork among small farmers along the middle Amazon floodplain has identified some promising strategies for agricultural development.

Forest Extraction

The value of plant products from forests of the Amazon floodplain has largely gone unappreciated, with potentially disastrous consequences. It has been suggested, for example, that the vocation for fertile alluvial soils in the humid tropics is agriculture rather than the harvesting of forest products (Pendelton 1992). Floodplain forests, it is argued, have fewer species than upland forests and, therefore, should act as a safety-valve for settlement and food production in order to "save" the biodiversity of the upland forests.

Floodplain forests may contain fewer species than upland forests because they occupy a much smaller area, and annual floods "screen out" many species not adapted to protracted inundation. Nevertheless, species richness of floodplain forests varies from one location to another, and some stretches of seasonally inundated forest doubtless contain more species of plants and animals than some upland forests. Furthermore, floodplain forests contain many endemic species. One cannot, therefore, afford to dispense with forests along the Amazon to make way for agriculture and other activities. The survival of forests is tied to efforts to intensify agriculture.

Inhabitants of the Amazon floodplain tap a large number of plant resources from flooded forests. Forest extraction for timber and non-timber forest products follows a seasonal rhythm. Some species fruit throughout the year, while others are ready for harvesting in the wild only during the rainy or dry season. Much of the logging for virola *(Virola surinamensis)* in the lower Amazon, for example, occurs at high water. Floodplain residents thus benefit from a year-round supply of food, building materials, and medicines gathered from the wild.

Several hundred plants in flooded forests are used by inhabitants along the Amazon River for a wide variety of uses (table 9.1). Forest extraction does not generate much income, nor is it likely to emerge as a mainstay for the household economy in most cases. Other than timber, only a few forest products from the Amazon floodplain are sold in significant quantities in urban markets. Nevertheless, floodplain forests are a significant supplement to the local diet and provide many useful products for the home.

Only a few of the extractive products from floodplain forests are reviewed here. Examples are drawn from a variety of uses, ranging from fruits, livestock feed, fish bait, construction supplies, and fire-wood. Floodplain forests serve as larder, fisherman's tackle box, living hardware store, and source of fuel to cook meals and prepare manioc flour.

Fruits for All Seasons

Floodplain forests provide a year-round harvest of fruits and nuts for inhabitants of the Amazon River. Palms, in particular, feature prominently in the diet of people along the banks and lake margins of the

Amazon. Açaí *(Euterpe oleracea)*, for example, provides fruits that are spun in simple, cottage industry machines to extract the pulp. The pulp is typically sold in plastic bags to be eaten fresh, particularly in the late afternoon, or is frozen for the ice cream and juice trade. Açaí juice is often sweetened with sugar and thickened with tapioca or manioc flour to form a porridge that sometimes becomes the last meal of the day. Some purists appreciate the savory flavor of açaí "straight."

Açaí occurs sporadically on the Amazon floodplain along the middle and upper stretches and is less important in the local diet and economy above the confluence of the Amazon and the Xingu rivers. In the Amazon estuary, however, graceful açaí forms dense groves where its purple or green fruits are gathered in large quantities for market(Anderson 1988, 1990). The collection of açaí fruits provides significant income for *ribeirinhos* living on alluvial islands of the lower Amazon. On Combu Island near Belém, for example, family income exceeds four thousand U.S. dollars per annum, most of it derived from selling açaí fruits (Anderson and Ioris 1992). Not all residents of the Amazon floodplain are blessed with such dense stands of economically important trees, nor are they always close to major cities, but the market may well accommodate the planting of açaí in cleared areas of the middle and upper Amazon.

Açaí is also harvested for its heart-of-palm (palmito), primarily for several canneries in the estuarine area (Brabo 1979). Some açaí heart-of-palm is exported but most is consumed by the well-to-do in urban areas of Brazil. If managed rationally, açaí groves can provide palmito on a sustained basis because they sprout when cut down. Selective pruning for heart-of-palm enhances fruit production (Anderson 1990). If, on the other hand, only the shoot is cut out, the palm dies. Some clandestine operators shimmy up açaí to extract the palmito, thereby destroying the resource. Conflicts sometimes arise between those who depend on açaí for fruits and those who harvest palmito.

The widespread buriti *(Mauritia flexuosa)* palm is especially concentrated along the upper Amazon, where large quantities of the pulp from its vitamin C–rich fruits are marketed in Iquitos (Padoch 1988). Deep-orange buriti juice and ice cream is a regional delicacy from Pucallpa on the Ucayali River to Belém at the mouth of the Amazon. In

Table 9.1. Some Local Uses of a Sample of Plants Found in Floodplain Forests Along the Amazon River

Local Name	Scientific Name	Use(s)	Source(s)
Abiu-rana	*Lucuma lasiocarpa*	Dark red wood used for furniture	Le Cointe 1947
Açaí	*Euterpe oleracea*	Fruit and heart-of-palm eaten; roots used to treat intestinal worms; inflorescences used as brooms	Anderson 1988, 1990; Anderson and Ioris 1992
Acapurana	*Campsiandra* sp.	Firewood	Bahri 1992
Assacu	*Hura crepitans*	Trunks used to make rafts	Bahri 1992
Buçú palm	*Manicaria saccifera*	Fronds used for house thatch along lower Amazon	Spruce 1908:59; fieldnotes (Macapá)
Buriti	*Mauritia flexuosa*	Fruit eaten	Padoch 1988
Catauari	*Crataeva benthami*	Fruit used as fishbait	Fieldnotes (nr. Santarém)
Castanha de macaco	*Couroupita guianensis*	Nuts fed to pigs and chickens	Fieldnotes (nr. Alenquer)
Espinheiro	*Acacia* sp.	Firewood	Bahri 1992
Envira	*Guatteria* spp.	Firewood	Bahri 1992
Farinha seca	*Licania micrantha*	Firewood	Bahri 1992
Fava rana	*Crudia amazonica*	Timber for building boats and canoes	Bahri 1992
Guaramã	*Ischnosiphon obliquus*	Stems used to make baskets and manioc sieves	Anderson and Ioris 1992 fieldnotes (nr. Santarém).
Ingá	*Inga* sp.	Firewood	Fieldnotes (nr. Santarém)
Envira	*Guatteria* spp.	Firewood	Bahri 1992
Jacaréuba	*Calophyllum brasiliensis*	Timber for building boats and canoes	Bahri 1992
Limorana	*Chlorophora tinctoria*	Firewood	Bahri 1992
Marajá	*Bactris* sp.	Ribs of fronds used as rafters for roofs; long thorns on trunk and leaves used as needles; fruit used as fish bait	Fieldnotes (Alenquer) Bahri 1992:67

Table 9.1. *continued*

Local Name	Scientific Name	Use(s)	Source(s)
Munguba	*Pseudobombax*	Inner bark used to tie fish together for market; rotten wood used for mulch	Smith, 1981 Fieldnotes (Ilha do Tapará)
Pau mulato	*Peltogyne paniculata*	Timber for construction, firewood	Fieldnotes (nr. Santarém)
Piranheira	*Piranhea trifoliata*	Trunks used as pillars for floodplain homes	Bahri 1992
Rubber	*Hevea brasiliensis*	Fruits used as fishbait; latex occasionally tapped	Fieldnotes (nr. Itacoatiara)
Sapucaia	*Lecythis pisonis*	Nuts eaten, also fed to chickens; inner bark used for cordage to tie fish	Fieldnotes (nr. Santarém) Smith 1981
Sapupira	*Diplotropis* sp.	Firewood	Fieldnotes (nr. Santarém)
Socoró	*Mouriria* cf. *ulei*	Fruits eaten raw and used as bait to catch tambaqui (*Colossoma macropomum*)	Smith 1981
Sumaúma	*Ceiba pentandra*	Trunks used for rafts; fiber around seeds used to stuff mattresses and pillows	Fieldnotes (Careiro Island)
Taperebá	*Spondias mombim*	Fruits eaten and used to make juice; inner bark used to treat diarrhea	Fieldnotes (nr. Macapá, Alenquer, Santarém)
Tarumã	*Vitex cymosa*	Fruit used as fishbait; timber for firewood	Fieldnotes (nr. Santarém)
Urucuri	*Attalea phalerata*	Fronds used in vegetable nurseries to protect seedlings; fruits fed to pigs Fruits burned to produce white smoke for coagulating latex of rubber	Bahri 1992 Fieldnotes (Urucurituba) Spruce 1908:185
Uruazeiro	*Cordia* sp.	Fruit used as fishbait	Fieldnotes (Obidos and Santarém area)

some areas of northern South America, heart-of-palm also is cut from buriti, and a sago-like starch is extracted from the pithy trunk to make bread.

In addition to açaí and buriti, floodplain forests harbor a number of other fruit trees that are much appreciated by locals as well as city folk. Several species of ingá produce pods with seeds surrounded by a sweet white pulp, and small quantities are sold in urban markets at high water, such as ingá baú *(Inga cinnamomea)* in Óbidos and the closely related marimari *(Cassia leiandra)* from Santarém upstream to at least Manacapuru. The more widespread yellow mombim *(Spondias mombim)* produces abundant orange-yellow fruits with an exquisite tart taste, when the waters begin to rise. Known locally as *taperebá* or *cajá*, the damson-sized fruits are typically mixed with water and sugar to make an enlivening drink. Modest quantities of yellow mombim fruits are sometimes taken to nearby urban markets where they sell briskly to make refreshing ice cream and sherbet. In some cases, yellow mombim densities on the floodplain provide bountiful harvests and significant quantities of the fruit reach urban markets for a month or so. At the main market in Macapá, for example, yellow mombim and watermelon were the most abundant fruits on sale in mid-December 1994. Other acidic fruits that are gathered from floodplain forests to make juice or to eat raw include several araças (species of *Psidium* and *Eugenia*), wild relatives of the cultivated guava.

Livestock Feed

A large number of floodplain plants contribute indirectly to human nutrition by providing food for cattle, pigs, chickens, ducks, goats, and turkeys. Virtually all residents of the Amazon floodplain keep some livestock, and even some small farmers are acquiring cattle. Feed and fodder for large and small domestic animals is thus a major preoccupation for floodplain inhabitants.

Several foods are gathered from floodplain forest to feed small livestock. The oily fruits of the ubiquitous urucuri palm *(Attalea phalerata)*, for example, are devoured by pigs. Urucuri palm grows on the higher parts of the Amazon floodplain and is often left standing in fields and near homes because of the useful fruits. Another floodplain palm, thorny jauarí *(Astrocaryum jauari)*, thrives in areas that stay under water longer, and its green fruits are collected to feed pigs, as is done near Nhamundá on the border between Pará and Amazonas states.

Capsules of the cannonball tree *(Couroupita guianensis)* are split open with an ax and the reddish brown pulp is fed to pigs and chickens. A relative of the Brazil nut tree, the cannonball tree produces neither sizable nuts for human consumption nor a pleasant-tasting pulp, but it is, nevertheless, often spared when people cut the forest for home sites or gardens. The majestic sapucaia *(Lecythis pisonis)* tree, another relative of the Brazil nut tree that grows in floodplain forests, produces nuts that are gathered for human consumption as well as for chickens and pigs.

Fishbait

The keen observational powers of fishermen have taught them that the fruits of numerous floodplain trees and bushes attract fish. In the vicinity of Itacoatiara and Santarém, fishermen comb floodplain forests for several fruits such as socoró *(Mouriria* cf. *ulei)*, tarumã *(Vitex cymosa)*, catauari *(Crataeva benthami)*, uruazeiro (*Cordia* sp.), and rubber *(Hevea brasiliensis)*, with which to bait their hooks. Fishermen on Careiro Island near Manaus (Bahri 1992:67) and along Paraná Nhamundá use the purple grape-sized fruits of marajá (*Bactris* sp.) for fish bait. One of the most prized fish, tambaqui *(Colossoma macropomum)*, is particularly fond of rubber seeds and the fruits of catauari and socoró. Shiny rubber seeds are gathered from the ground in large quantities for sale to urban fishermen and, in the past, served as a snack food for people. Riverine inhabitants also appreciate the ripe red fruits of socoró.

Aracu (various species of *Leporinus*, *Rhytiodus*, and *Schizodon*) are reputed to be especially fond of uruazeiro. Striped aracu do not command the same high price in fish markets as tambaqui but are nevertheless appreciated for domestic consumption and by poorer urban residents who can better afford their lower market price. Fishermen in the vicinity of Santarém also claim that tarumã fruits are particularly useful for catching pirapitinga *(Colossoma bidens)*, related to tambaqui but smaller, and jatuarana *(Brycon* sp.), both fish with ready markets.

Construction Supplies

Even the simplest homes on the floodplain require some wood, if only for posts. A favored tree for such purposes is pau mulato *(Peltogyne paniculata)*, now increasingly rare on the floodplain to which it is confined. In the nineteenth century pau mulato was abundant along the Amazon up to the foothills of the Andes (Spruce 1908:153). Pau mulato is easily spotted by loggers because of its distinctive smooth red trunk and branches. Piranheira *(Piranhea trifoliata)* is the preferred wood for stilts to raise house floors above floods. Piranheira's extremely hard wood resists termites and prolonged waterlogging. But it too has become difficult to find because of rampant deforestation and uncontrolled logging, such as on Careiro Island near Manaus (Bahri 1992). Along the Upper Amazon in Peru, *ribereños* use rot-resistant huacapú *(Vouacapoua americana)* for the main posts of their homes (Hiraoka 1985).

Many homes along the Amazon floodplain are walled and covered with thatch and must be substantially rebuilt after each flood. The generous, arching fronds of buçú *(Manicaria saccifera)* are favored for roofing homes in the Amazon estuary where the palm abounds. Durable buçú palms also are marketed in towns in the lower Amazon, as observed in Macapá in December 1994 and Moju, Pará, in March 1992. Rafters of palm-thatched homes are sometimes made from the stiff mid-ribs from fronds of the marajá palm (*Bactris* sp.). Marajá palms tend to occur in clumps in the understory of floodplain forest; the impressive black spines that protrude from the trunk and underneath the fronds serve as makeshift needles.

Some floodplain residents construct floating stores and corrals for their cattle. Massive logs of the kapok *(Ceiba pentandra)* tree are strapped or nailed together with planks to form a raft, upon which stores or corrals are built. Now such sights are increasingly rare. Most of the sizable kapok trees, known as sumaúma in the Brazilian Amazon, have been cut down to float hardwoods along the Amazon or to provide inexpensive core wood for veneer plants. Another tree of floodplain forests used for such rafts is assacu *(Hura crepitans)*. Ever increasing numbers of farmers and ranchers are opting to rent or buy upland pasture, a process that accelerates deforestation on the uplands.

Fuelwood

In rural areas of the Amazon, wood and charcoal are the primary cooking fuels. The search for wood to prepare meals is thus a daily chore for most residents of the Amazon floodplain. It is easy to conceive of fuelwood shortages in dry regions, but more difficult to imagine a fuelwood crisis looming in the well-watered Amazon Basin. Yet as floodplain forests along the Amazon continue to shrink, fuelwood supplies are likely to emerge as a resource issue.

Fuelwood is gathered from floodplain forests, from driftwood left by receding floods, and to a lesser extent from home gardens. Not all woods have the same heating value, however, and certain species are preferred for cooking and to prepare manioc flour. Sought-after woods for roasting, boiling, or frying food in the Santarém area include pau mulato, ingá, and sapupira (*Diplotropis* sp.). Few people bake bread along the Amazon, but many are accustomed to eating manioc flour. Preparation of manioc flour is a time-consuming process and the heating of the gritty flour on a metal griddle takes several hours. Because of the prolonged baking time, slow-burning wood with a high heat value is preferred. On Careiro Island near Manaus, such woods include envira (*Guatteria* spp.), limorana *(Chlorophora tinctoria)*, espinheiro (*Acacia* sp.), farinha seca *(Licania micrantha)*, acapurana (*Campsiandra* sp.), tarumã, and piranheira (Bahri 1992:67). Many floodplain residents make manioc flour at low water to sell in nearby towns and cities. Plentiful supplies of fuelwood are essential to their livelihood.

Conservation of fisheries should be central to any overall plan to develop and manage natural resources on the Amazon floodplain. But little attention has been paid to how future generations on the floodplain will cook their fish. The importance of fuelwood to the well-being of floodplain residents again underscores the imperative task of wisely protecting and managing forests along the Amazon.

The Pivotal Role of Home Gardens

At first glance, home gardens may seem an unlikely place to search for clues to devise commercial agroforestry systems for the Amazon floodplain. Home gardens are commonly perceived as backyard collections of plants for purely domestic use with little economic value. Home gardens are a form of agroforestry, however, albeit limited in scale. In most cases they contain too many species for implementation in large fields, but the species richness of home gardens is an asset when looking for promising crops for more widespread planting. Furthermore, home gardens are much more than mere supplements to the diet and shady fora for afternoon discussions. Home gardens on the Amazon floodplain and elsewhere in the region are dynamic staging grounds for launching new crops. They serve as living gene banks for improving agriculture (Smith 1996).

Home gardens on the Amazon floodplain are relevant to agroforestry development on four counts. First, wild species from the forest and other environments on the Amazon floodplain are being actively domesticated for a wide variety of purposes. Some of these species could in the future be major crops. Second, home gardens are low-cost arenas for trying out new crops from other regions. If the introduced plants perform well, they may be cultivated on a larger scale. Home gardens are thus an integral part of the risk-aversion strategy of small farmers. Third, the partial shade of home gardens is ideal for establishing seedlings, such as cacao and cupuaçu *(Theobroma grandiflorum)*, destined for transplanting into groves or fields. Fourth, home gardens are important depositories of varieties of crops adapted to occasional flooding, some of which could prove suitable for more widespread planting.

One of the common misconceptions about the Amazon floodplain is that periodic inundations negate the possibilities of perennial cropping. It is commonly assumed that alluvial soils are best used for annual crops or pasture. But the formerly extensive floodplain forests are testament that many woody species are adapted to varying degrees of flooding. Furthermore, residents of the Amazon floodplain, such as on Careiro Island, have long cultivated a mix of indigenous and exotic crops in their home gardens, some of which have provided significant income (Sternberg 1995).

Home gardens on the Amazon floodplain contain a surprising number of plants. Over eighty species of perennial plants alone are currently cultivated in backyard gardens in the stretch of the Amazon floodplain from Juriti to Santarém. In surveys of home gardens on uplands and the Amazon floodplain, species composition differs between the two environments, but no significant difference in species richness was noted. Floodplain home gardens are likely to prove fertile hunting grounds for preadapted genotypes of mango, cupuaçu, banana, guava, cashew, and many other fruit and nut trees that can be grown on a commercial scale.

Commercial Agroforestry

Although the notion of commercial agroforestry for the Amazon floodplain may seem a radical proposition, it is not a new idea. In the eighteenth century cacao was planted extensively along the lower Amazon, under royal edict. In the latter part of the nineteenth century and in the early twentieth century, the rubber boom spurred extensive planting of *Hevea brasiliensis* on the Amazon floodplain. Rubber and cacao also may have been native to certain parts of the Amazon floodplain; human agency has enriched forests with these economic species. Until the mid–twentieth century, cacao and rubber were still commercially significant crops along the middle and lower Amazon. In some areas of the Amazon floodplain, such as Careiro Island near Manaus, cacao and rubber groves were enriched with various fruit trees to form multitiered orchards (Bahri 1992).

Many of the agroforestry systems on the Amazon floodplain have fallen into disrepair or have vanished. The unusually severe flood of 1953 is often cited by farmers between Nhamundá and Santarém as a triggering factor in their decision to abandon cacao cultivation. Falling prices of cacao and rubber in the 1980s are another reason these crops are now less common. The jute boom of the middle of this century, coupled with the recent expansion of cattle and water buffalo ranching, have laid waste to many intricate and formerly extensive agroforestry plots on the Amazon floodplain. Only in the Amazon estuary area, particularly close to Belém, are extensive cultural forests providing the main livelihood for rural folk. On Combu Island, for example, polycultural orchards containing a mix of wild and cultivated species have replaced slash-and-burn agriculture as the mainstay for farmers.

What is the potential for agroforestry for less privileged locations of the Amazon floodplain? No single model or agroforestry configuration is appropriate for the entire stretch of the river. Environmental heterogeneity and differing cultural and market conditions preclude any one blueprint for agroforestry development. By working closely with farmers and analyzing market conditions and socioeconomic constraints, however, it should be possible to promote profitable agroforestry systems on already deforested areas of the Amazon floodplain.

The Amazon floodplain has three major advantages over upland areas for commercial agroforestry. First, soils are generally fertile and are annually renewed by floods. Fertilizer costs would, therefore, be low compared to the generally leached soils of uplands. Second, water transportation is much cheaper than by road. In the Santarém area, for example, river transport is about one-half the cost of truck freight charges for communities inland from the city. Third, most of the people in the region live along the Amazon River, or in close proximity to it.

The Amazon floodplain has been virtually ignored by development planners for several decades. Brazil's cacao research and development agency, Comissão Executiva do Plano da Lavoura Cacaueira (CEPLAC), has focused its research and development efforts in Amazonia on the uplands, for example, particularly areas with soils belonging to the alfisol group, such as those near Altamira in Pará and in Rondônia. Given the paucity of information about plant resources on the Amazon floodplain, a wise course would be to concentrate initially on a few relatively well known crops with secure markets.

Among the crops with ready markets in Amazonia are cupuaçu, mango, and banana. All three crops are found in home gardens on the floodplain. Cupuaçu tolerates moderate flooding and thrives on the higher parts of the floodplain. Grafted selections of mango with high market value (such as Haden and Keitt) could be planted either in agroforestry plots or monocultural orchards. Banana is one of the most important perennial crops on the middle Amazon, where it is usually planted in monocrop stands, but it could also be planted in agroforestry plots. Banana is well adapted to floods and produces fruit in home gardens at high water. Prolonged inundation kills banana, but only temporarily, since it puts forward new shoots when the waters recede. Floods assist banana cultivators by destroying some pests and weeds and by replenishing the soil.

A parallel effort should be undertaken to assess the potential of less well known crops and plants currently in various stages of domestication in home gardens. Various unusual fruit trees are cultivated in home gardens or are found in the wild. Some of these could eventually be developed into commercial propositions. But the process of selecting promising genotypes and securing a market for their products is time-consuming and costly. For this reason, a substantial part of the research and development effort needs to focus on conventional crops to improve the chances of an early payoff.

Several forest fruit trees are deliberately planted or spared in home gardens because they are useful as fish bait. As aquaculture becomes more common in upland areas, several of these species could be planted around ponds to enrich the diet and improve the flavor of such fruit-eating fish as tambaqui, one of the most commercially valuable fish in the Amazon. Fish-bait trees also could be planted in orchards on the Amazon floodplain, and the fruits sold to fish farmers in upland areas. On the Amazon floodplain, fish-bait trees could be planted around certain lakes to promote fisheries. Forests and home gardens on the Amazon floodplain could thus benefit entrepreneurs and consumers in other parts of the basin or even other regions.

Another possibility for commercial agroforestry on the Amazon floodplain is timber production. In some upland areas of the Amazon, farmers are incorporating several timber species as part of their agroforestry systems (Smith et al. 1995). Several commercially valuable timber species also are adapted to the Amazon floodplain, but most have been heavily logged since they are relatively accessible.

Timber production on the Amazon floodplain is warranted to meet domestic needs and to generate income. Residents of the Amazon floodplain seek hardwoods to build homes, boats, and canoes. In the past several floodplain forest trees such as fava rana *(Crudia amazonica)* and jacaréuba *(Calophyllum brasiliensis)* were used extensively for such purposes (Bahri 1992:66). Now, with so many of the floodplain forests gone or depleted along the Amazon, craftsmen have been forced to buy durable wood from suppliers along the Upper Amazon or in the uplands. In the middle Amazon, at least, most boats and canoes are now built with the deep red wood of pau d'arco (*Tabebuia* spp.) or tan-colored itaúba *(Mezilaurus itauba)*, both giants of terra firma forests.

As development pressures sweep upstream and upland forests are "high-graded" of their valuable timber, such options will become more difficult. Timber production on the Amazon floodplain could be feasible as part of agroforestry plots owned by small farmers, on larger plantations, or from managed forests. Little, if any, thought has been given to the rational management of floodplain forests for timber and other products or to reforestation along the Amazon. Now is the time to take action on conservation of these resources, before they disappear to the detriment of floodplain residents.

A Way Forward

In order to safeguard and better utilize plant resources along the Amazon floodplain, several conservation and development issues must be tackled. Different segments of the Amazon River should be preserved to serve as food and shelter for many fish species, as genetic reservoirs, and as sources of numerous products for local people. Two main types of reserves are needed: nature preserves with little or no human disturbance; and multiple-use reserves where certain animal and plant products are harvested without seriously affecting the integrity of the ecosystems involved.

Setting up a reserve in only one location, even if it is large, will not suffice. The marked ecological heterogeneity of the Amazon floodplain necessitates a multilocational approach to both conservation and development. Two sizable reserves with varying degrees of protection for nature have been established at the confluence of the Ucayali and Marañon rivers in the western part of the Amazon basin and at Mamirauá near Tefé, but many other biologically interesting areas of the Amazon floodplain lack formal protection or lack efforts to promote the sustainable harvesting of their natural resources. Much work remains to be done to identify high priority areas for conservation on biological grounds, as well as those that might be feasible as protected or managed areas from cultural and political perspectives.

To promote agroforestry, several steps are called for. Promising crops must be multiplied in well-run nurseries and then distributed to farmers. Both the private and the public sectors can usefully be employed in this process, but some credit facilities will be needed. The land tenure situation on the Amazon floodplain is much less contentious than in some other parts of the Amazon basin because traditional landownership patterns have not been challenged by a large influx of migrants and speculators. Land is changing hands on the Amazon floodplain, but in a peaceful manner. Still, many small farmers lack formal documents, and official land titles will need to be expedited—or other legal mechanisms devised—to allow banks to make loans to small farmers. More support is warranted to launch agroindustries to process tree crops from the Amazon floodplain, agroindustries such as frozen fruit juices, fruit-flavored yogurts, essential oils, and other value-added processing. Some agroindustrial plants could be assembled on barges so they can be towed to areas where there is a high demand for their services, as in the case of some fish-freezing and ice-making plants.

Successful policies to promote agroforestry and the conservation of plant genetic resources on the Amazon floodplain hinge on blending indigenous knowledge and scientific expertise as well as the involvement of all stakeholders. Policies that do not take into account large landowners and other land-use systems on the Amazon floodplain are likely to fail. Currently, much of the development debate in Amazonia focuses on community management of resources. Excessive reliance on this approach is likely to backfire. The sense of community is often weakly developed on the Amazon floodplain, and one senses that attempts to foster community spirit are often instigated by outside groups rather than by internal, grassroots efforts. One cannot even count on religious homogeneity in rural populations to help encourage community cohesiveness; many nominal Catholics have converted to various Protestant denominations, especially Pentecostal groups. As in the case of other stakeholders, communities may be motivated to cut down forests for economic reasons. Certainly, communities that seek outside assistance to manage their natural resources deserve support, including training. A mosaic of approaches to conservation and development in the Amazon is needed, ranging from community-based management systems, where appropriate, to individual smallholders and larger landowners.

The private sector is critical not only for conservation of floodplain forest but also for the development of some tree plantations and agroindustry. Much of the remaining forest along the middle Amazon is on ranchland, and what motivates ranchers to hold on to forest can teach us valuable lessons. Furthermore, agroforestry will not replace cattle ranching on the Amazon floodplain, at least not in the near future. Cattle and water buffalo provide many useful products for small farmers, such as fertilizer for vegetable crops. How to incorporate ranchers and individual farmers in conservation and NRM plans represents a major challenge. But with sufficient knowledge and the inclusion of all actors on the Amazon floodplain, agroforestry can regain lost ground and eventually become an important source of income for both small- and large-scale operators.

Notes
Bibliography
Index

Notes

Introduction

1. By "risk-prone environments" is meant those environmental zones that do not readily respond to intensification efforts to increase yields. The number of persons living in these zones in Latin America alone is estimated to be about one hundred million.

2. As used here, the word "traditional" does not imply a static, backward-looking system of knowledge. All "traditional" NRM systems are evolving in response to changes in the environment, including market opportunities. Traditional farmers include small, peasant, and indigenous farmers.

3. It would be worthwhile to the themes addressed in this volume to attempt some statistical analysis of project "success" based on level of participation by target beneficiaries. However, such an exercise faces severe limitations because real data are lacking. As Quiroz remarks in an endnote, even a general overview is difficult because "not all of the projects related to local knowledge systems in Latin America have published results (such as many grassroots efforts) and that even the CIKARD documentation unit still lacks reports from some of those that have been published."

4. For a complementary study with listings of floodplain floral and faunal resources utilized by local peoples, see Frechione, Posey, and Francelino da Silva (1989).

1. Participatory Technology Development in Risk-Prone Areas

1. Although the term "indigenous knowledge systems" has become standardized in the literature, in this chapter we use the terms "traditional," "local," and "indigenous knowledge systems" to refer to the understandings of the world and the workings of the natural environment as developed by rural

peoples from their observations of their immediate surroundings. Traditional systems encompass the combination of knowledge, productive resources, input, and services that are applied systematically to produce desired outputs. They include both the physical forms of technology (such as tools, seeds, and so on) and the methods, practices, and strategies (including forms of social organization). "Rural peoples" include not only "native peoples" but also peasant farmers, settlers, and other resource users.

2. These differences are recognized in a recent World Bank (1993a) document delineating a strategy for the agricultural sector in the LAC region.

3. Contribution of the agricultural sector to GDP is 10 percent in LAC, compared with 25 percent in the Europe, Middle East, and North Africa region (EMENA) and East Asia, and 30 percent in Africa and South Asia. The percentage of rural population is 30 percent of the total population in LAC, compared with 50 percent in EMENA and East Asia, and 70 percent in Africa and South Asia.

4. For example, Brazil and Mexico together account for 51 percent of the area of LAC, 54 percent of the population, and 58 percent of the region's agricultural GDP. These two countries with Colombia and Argentina account for 71 percent of the area, 69 percent of the population, and 77 percent of the region's agricultural GDP.

5. For example, in Brazil, agriculture still contributes 40 percent of export receipts and 25 percent of employment.

6. For example, in Colombia, with a population less than 42 percent rural, 74 percent of the poor live in rural areas. In Brazil, where nearly 26 percent of the population is rural, over 40 percent of the poverty is rural. In Venezuela, where 16 percent of the population is rural, 30 percent of the poor is rural (Wiens 1994).

7. Performance of the agricultural sector has been quite variable among LAC countries. For example, the regional rate of growth of agriculture during the 1980s was 1.9 percent per year. However, eight countries experienced agricultural growth of over 2.5 percent, with six over 3 percent (among them, Ecuador and Chile grew 4 percent); four countries grew at between 1 and 2.5 percent; and nine others grew at less than 1 percent, with four recording negative agricultural growth during the decade. Although agricultural growth rates exceeded the growth rates of total GDP in a majority of LAC countries, the relationship was reversed in the last half of the period between 1980 and 1993 (Wiens 1994). The largest LAC countries are now undergoing forced structural adjustments in the agricultural sector, which, in the short term, can have negative impacts on growth.

8. Ecuador is a puzzling case because of its highly uneven subsectoral growth. It was the fastest-growing agricultural economy in the 1980s (at 4.3 percent per year), primarily on the basis of shrimp farming and banana production (both for export). By contrast, growth was very slow in other subsectors directed to domestic markets located in the highlands, despite considerable public expenditure (World Bank 1993a).

9. Although the share of the national workforce employed in agriculture declines with economic growth, it declines at a much slower rate than agriculture's share in national income. Moreover, the absolute number of agricultural workers typically does not peak until a country attains upper-middle-income status (World Bank 1993a). Thus, as countries develop, agriculture faces the double challenge of how to absorb additional workers and how to increase the productivity of the average worker to reduce poverty (World Bank 1993a).

10. Indeed, UNCED's Agenda 21, adopted in 1992, reflects global recognition of the interrelated problems of poverty, low agricultural productivity, and resource degradation. It also has elicited a global commitment to combat poverty, while ensuring sustainable agricultural and rural development. The main instruments to achieve those objectives outlined under Agenda 21 include policy and agrarian reform, people's participation, income diversification, infrastructure and human resource development, land resource conservation and rehabilitation, and improved management of inputs.

11. Wiens (1994) reports that there also is a close coincidence between ethnicity and rural poverty in risk-prone areas. Indigenous people are usually landless farm laborers in commercial farming areas or smallholders in areas of marginal quality with a low degree of commercialization. A recent World Bank study estimates the indigenous population of LAC at between nineteen and thirty-four million people, with the great majority living in Bolivia, Ecuador, Guatemala, Mexico, and Peru (Psacharopoulos and Patrinos 1994). The same study calculates that 80 percent of the indigenous population is poor (if lower estimates of the total population are used) and that over one-half are extremely poor.

12. See Bilsborrow (1992) for an excellent discussion.

13. Referring to the effects of population growth on agriculture and technological innovation, Boserup (1965, 1981) challenged the Ricardian-Malthusian assumption of constant technology, postulating that as rural land becomes scarce relative to population (because of population growth), land will be used more intensively in order to produce greater yields. Boserup sees agricultural intensification as an ecological imperative based on a new human/land ratio: as long as the number of people continues to rise on a given amount of land,

or the amount of land available continues to fall for a constant number of people, the need for increasing labor investment is expected to continue as well. She then proposed (Boserup 1965:56) that "the growth of population is a major determinant of technological change in agriculture," which stems from the necessity of raising output per unit of area to offset the increase in labor requirements associated with more intensive land use. According to Boserup, population becomes the independent variable, while the dependent variables become agrotechnology, the intensiveness with which labor inputs are applied, and hence, the capacity of the system to support people.

14. Instead of promoting the intensification of agriculture (through labor-intensive land investments, higher output per unit of area, or improved management of areas already under cultivation), governmental policies have often relied on frontier expansion or land extensification (Bilsborrow and Ogendo 1992), usually with disastrous environmental consequences. Other misguided public policies that have contributed to the loss of forest cover in large areas of LAC include (1) open access, leading to waste and misuse of forest resources; (2) deforestation as a requisite for land tenure; (3) tenure insecurity; (4) the demise of common property regimes; and (5) government tax exemptions and credit incentives for pastures, cattle ranching, and short cycle crops (Binswanger 1987; Mahar 1988; Poveda 1991; Southgate and Runge 1990; Uquillas 1984).

15. For example, coffee processing has had serious effects on lands and waters, contributing to about two-thirds of all water pollution in Costa Rica. In Honduras, the greatest threat to the Gulf of Fonseca is water pollution caused by the misuse of pesticides (DeWalt, Vergne, and Hardin 1993). Worldwide, some twenty thousand deaths and one million illnesses each year are attributable to pesticide poisoning—most of these cases occur in developing countries. Expansion of irrigated agriculture also can lead to land degradation through waterlogging and salt accumulation as a result of poor drainage and excessive use of water. About 24 percent of the world's irrigated land is affected by salinization to varying degrees (Pinstrup-Andersen, Pandya-Lorch, and Hazell 1994).

16. The data are taken from various standard international sources: FAO (1992, 1993); IUCN/UNEP/WWF (1991); UN (1984, 1985, 1989); WCMC (1992); World Bank (1990, 1991, 1992, 1994); WRI (1989, 1990, 1992); WRI/ CIDE (1990); WRI/CIDE and USAID/LAC (1991); and Zachariah and Vu (1988). Most of these sources, in turn, ultimately derive their data from the countries themselves. The data are evidently of varying quality and are used here mainly to illustrate general patterns.

17. Although serious problems of watershed destruction and subsequent soil erosion have resulted from the deforestation of highland forests, the bulk of the literature refers only to the depletion of lowland forests or tropical moist forests and the consequences in terms of loss of biodiversity.

18. There are no comprehensive studies of the region that show the magnitude of the desertification problem in LAC. However, according to CEPAL (1989), about 18 percent of South America is estimated to have a moderate or high degree of desertification hazard.

19. However, the spread of modern varieties of pulses, oilseeds, vegetables, and other basic food crops such as sorghum, millet, and root crops has been very limited.

20. Hayami and Ruttan (1987) have argued against the view that green revolution technologies have generated inequity in rural incomes. Although they recognize that, in some cases, small or poor farmers lagged significantly behind the large or wealthy farmers in the adoption of green revolution technologies, they suggest that such cases are largely a reflection of institutional rather than technical bias. These authors argue that institutional reforms are necessary to partition equitably the new income streams generated by an appropriate technology. Kuznets (1980) also has emphasized that a successfully developed country has to acquire the capacity both to encourage technological innovation and to design institutions that can accommodate the uneven distributional impact of technology on different social groups.

21. Formal scientific research has received much attention and resources, and as a result, progressive "counter-ideologies" have tended to romanticize the role of farmers as sources of innovation. This may create exaggerated expectations, as if farmers alone held the key to solving problems that a science-based approach has failed to solve. It is thus not surprising that the indigenous knowledge literature has often been ignored by agricultural scientists because of the sometimes missionary fervor with which proponents preach the virtues of indigenous knowledge systems (DeWalt 1994). Neither stance is helpful. The crucial question is, How can the various actors be brought closer together in a well-functioning system of generating agricultural knowledge and technology in which their activities complement and reinforce each other?

22. In their constant struggle to sustain their agricultural livelihoods, resource-poor households and communities in risk-prone environments have developed innumerable ways of obtaining food and fibre from the natural environment. Many traditional farming systems were sustainable for centuries in terms of their ability to maintain a continuing, stable level of production (TAC/CGIAR 1988). A wide range of different farming and animal husbandry

systems evolved, each developed to adapt to local ecological conditions and inextricably entwined with local culture. A closer examination of several of these traditional NRM systems reveals they are not static and they have, in fact, changed over the generations—and particularly quickly over the last few decades. Rapid changes in demographic, economic, and technological conditions have caused rapid transformations in traditional systems. Increasing population pressures, market integration, indiscriminate promotion of modern inputs, and financial constraints have all led farmers and other natural resource users to seek short-term profits and to neglect investments in the long-term sustainability of their natural resource base.

23. Scientists and farmers are only two groups of actors in a complex system of generating knowledge and technology for agriculture. Besides the scientists working in basic, applied, and adaptive research, many extensionists and other development workers in national agencies and in bilateral and multilateral government projects also are involved in experimentation and adaptation of technology to local conditions (Farrington and Martin 1988). NGOs and farmer organizations are likewise involved in the development of agricultural technology. Artisans and traders in the informal sector and commercial firms in the formal sector of the national economy also seek to develop and supply agricultural inputs and to develop ways of processing and distributing agricultural products.

24. Among developing countries and donor agencies there is growing interest in understanding the role that indigenous knowledge can play in making development projects more effective and efficient (Barborak and Green 1987; Bheenick, E. G. Bonkounguo, and others 1989; Franke and Chasin 1980; Hoskins 1984; Verhelst 1990). Specific examples are found in Niamir (1990), Titilola (1990), Dommen (1988, 1989), Cashman (1989), Poffenberger (1990), Norem, Yoder, and Martin (1989), Ascher and Healy (1990), Jodha (1990), Warren (1991a), Warren and Cashman (1989), Butler and Waud (1990), Cernea, Coulter, and Russell (1985), DenBiggelaar (1991), Denning (1985), Fairhead (1990), Moris (1991), Rolling and Engel (1989), Cook and Grut (1989), Groenfeldt (1991), Messerschmidt (1991), Rau (1991), and Uphoff (1985).

25. For example, the Incas created a sophisticated system of irrigation and canals that enabled communities to utilize water and other natural resources with remarkable efficiency. This heritage was destroyed by demographic, economic, and political changes imposed on native people during the colonial (1532–1821) and republican (after 1821) periods. One of the main centers of Mayan cultivation also seems to have failed, after centuries of expansion in Guatemala, because the demands of its growing population depleted the soil through erosion.

26. Of the total value of seeds used worldwide (circa fifty billion U.S. dollars in the mid-1980s), farmer-saved seeds account for about 35 percent (Jaffee and Srivastava 1992).

27. Numerous NGOs are working in risk-prone areas. According to Chambers, Pacey, and Thrupp (1989), these organizations have tended to be strong in their links with grassroot communities but weaker on the technical side of farming. Nevertheless, several NGOs such as World Neighbors, Winrock, Oxfam, and IIRR have successfully worked with smallholders to develop appropriate agricultural technologies.

28. Examples of detailed economic analyses include a long-term case study in Western Honduras by Felber and Foletti (1989), which found that green revolution corn technology offered a lower economic return than traditional growing practices. Mausolff and Farber (1995) also compared the economic costs and benefits of chemical and low-purchased-input ecological technologies in two Honduran rural development projects and found that traditional practices based on cover cropping with velvet bean *(Macuna pruriens)* have tripled average corn yields from a baseline of between seven hundred and two thousand kilograms per hectare, using only one-fifth of the commonly applied chemical fertilizer.

29. Where the key to creditworthiness was perceived to lie in collateral, land-titling projects or components were often undertaken to tackle the problem. However, a recent review of the project record (Wachter and English 1992) indicates that performance fell far short of the objectives. It is now recognized that the key is not collateral per se, but a proved track record of repaying debt. There are examples in almost every country in the LAC region of agricultural credit projects that failed to redistribute income toward low-income agricultural producers despite subsidized interest rates for such borrowers. Even in the most favorable cases, only a small proportion of low-income producers ever receive low-interest project loans, while a substantial share of the low-interest credit provided under these projects seems inevitably to flow to relatively high-income producers (World Bank 1989).

30. All references to dollars are to U.S. dollars.

31. Quite apart from formidable administrative problems, there are other powerful economic and political forces to overcome. In countries with highly uneven land distribution, however, the potential benefits—in terms of employment-intensive agricultural growth, social stability, and an improved climate for private investment—may be so powerful that market-assisted land reform should be tested (World Bank 1993b). Failure to reform highly dualistic systems of landownership has often resulted in peasant uprisings and sometimes decade-long civil wars, as in parts of Central America.

32. The Bank also is completing major analytical work on three different issues: (1) the case for nonmarket, long-term finance for forestry; (2) the content of promotion of efficient land markets; and (3) the utility of land zoning as an approach to sustainable natural resources management in settled, nonfrontier areas. Other questions being addressed include how to manage the provision of rural infrastructure; how to make efficient use of the large network of public agricultural education institutions; and what, precisely, the Bank should do in irrigation (World Bank 1993a).

33. Wiens (1994) argues that irrigation and water rights are as unequally distributed as landownership in Latin America, and that the effects on efficiency may be as serious as the equity issue. Restoring water rights usurped from the highlands in the Andes, for example, and rehabilitating traditional conservation structures and management institutions could increase farm productivity by an estimated 40 percent.

2. Rural Development and Indigenous Resources

1. Such cross-cultural communication entails collaboration with local people in mapping the environment in effective and meaningful ways. This is an important and newly emerging aspect of geographic-based assessment-framework development. For detailed information on this approach, see the special issue of *Cultural Survival Quarterly* on *Geomatics: Who Needs It?*—especially the article by Wilcox and Duin, "Indigenous Cultural and Biological Diversity: Overlapping Values of Latin America Ecoregions," *Cultural Survival Quarterly* 18.4 (1995):49–53.

2. For the purposes of this discussion, the term "smallholder development" is taken to encompass poverty alleviation and general improvement of human health and welfare and environmental quality. It is assumed to include the maintenance and enhancement of ecological health and biotic integrity (see Karr 1995).

3. Concepts of capital and substitutability applied to natural resources and biological diversity have been defined and discussed by Costanza and Daly (1992) and others. The definition of indigenous resources expressed here is consistent with these concepts, those relating to the concept of contributory value of biological diversity (Wilcox 1994), and the notion of "indigenous" as applied to biological species.

4. "Indigenous" need not be restricted to native or traditional cultures per se but may include any people who have gained local knowledge—regardless of ethnic background, origin, or length of occupancy of the land.

5. A significant proportion of the LAC region is not humid tropical but the areas lying outside the humid tropical zone, which are more suitable for conventional intensification (with rich soil and water and access to markets), have been largely co-opted for industrial-scale production.

3. Combining Indigenous and Scientific Knowledge

1. I prefer the term "local knowledge systems" to "indigenous knowledge systems," because "indigenous knowledge" carries the connotations of native peoples' ideas and beliefs and of traditional knowledge. Nonindigenous people, however, also have developed understandings of the world that are based on their observations of their immediate surroundings. These understandings are constantly being modified as people seek adaptive strategies that are more effective in their particular ecological, social, political, and economic contexts. It is these local understandings that we are trying to capture through the study of their knowledge systems.

Despite my preference, the term "indigenous knowledge systems" has become standardized in the literature. A substantial number of publications, national and regional indigenous knowledge resource centers, and an international newsletter *(Indigenous Knowledge and Development Monitor)* all use the term. For this reason, with the proviso that we mean *all* local knowledge systems, I retain the term here as defined by McClure (1989:1): "Indigenous knowledge systems are learned ways of knowing and looking at the world. They have evolved from years of experience and trial-and-error problem solving by groups of people working to meet the challenges they face in their local environments, drawing upon the resources they have at hand."

2. Latour (1986) attributes the great power of science to what he calls "inscriptions"—the ability to produce images and to read and write about them. Because these images can be superimposed, reshuffled, recombined, and summarized and because they are communicable to others, they acquire substantial importance. "By working on papers alone, on fragile inscriptions which are immensely less than the things from which they are extracted, it is still possible to dominate all things, and all people. What is insignificant for all other cultures becomes the most significant, the only significant aspect of reality" (Latour 1986:32).

3. A prime example here might be the search for hybrid seeds that satisfy the goals of increased productivity but whose organoleptic qualities may result in less desirable food.

4. Many individuals talk about indigenous knowledge as though it were a

highly codified system. Indigenous knowledge, however, is very unevenly distributed among the individuals who make up communities. There are exceptionally knowledgeable individuals and there are often "specialists" who have a great deal of knowledge of certain realms. Identifying these specialists or gifted informants is an important first step in learning about local knowledge.

5. It is difficult to construct a synthetic view of a typical indigenous knowledge system. This is because any group of people in any kind of setting may share derived wisdom about agriculture, natural resources, or other knowledge domains. These are all local or indigenous. Even farmers who are part of modern industrial agriculture have an indigenous knowledge system. In this section, I am mainly contrasting characteristics of societies that have had little contact with scientific knowledge systems with those that depend primarily on science for the construction of their knowledge systems.

6. Jared Diamond (1987:64) has opined that the adoption of agriculture may have been "the worst mistake in the history of the human race." He reports, "Forced to choose between limiting population or trying to increase food production, we chose the latter and ended up with starvation, warfare, and tyranny" (66). Instead of producing a better life, the agricultural revolution was "a catastrophe from which we have never recovered. With agriculture came the gross social and sexual inequality, the disease and despotism, that curse our existence" (64).

7. Researchers are not given the same kind of academic rewards or other recognition for applied research as they are for basic research. One of the trends that Busch and Lacy (1983) have documented for all the agricultural disciplines is that over time each has become more like a basic "scientific" discipline, with journals and academic meetings that would be incomprehensible to any lay person.

8. Even plant breeders who create and re-create whole plants are somewhat specialized in these terms, because they may not be cognizant of the cropping system into which their particular species will fit. One of the unfortunate consequences of this has been the trend toward monoculture, where an "ideal" environment is provided in order to derive maximum productivity from that one species.

9. It is important, however, that we recognize that establishing mutual respect between scientists and producers of local knowledge will not necessarily create more just and ecologically sustainable systems. Local farmers with intimate and intricate ecological understandings have not been immune to the destruction of their own ecosystems (Eckholm 1976). In the same vein, farmers

are not above exploiting their neighbors, so that the technology they create or participate in developing will not necessarily create a more just socioeconomic system (Flora 1992:96). Within the process of cultural evolution, technology and science are only parts of our society's adaptive strategies; whether they fit with our goals for society or whether they are ecologically sustainable is a different set of issues. Means for accomplishing these ends should become part of the system of scientific investigation and technology development (DeWalt 1991b).

4. Organizing for Change—Organizing for Modernization?

1. In the current economic policy context, the other sustainable NRM option is to follow the path of the U.S. eastern seaboard in previous centuries, a process whereby progressive out-migration from rural areas to cities and a shift from rural to urban livelihoods led to reduced pressure on local natural resources, low-intensity rural resource management systems, and the return of secondary forest over massive areas of the coastal states. This option depended, however, on a steady and rapid expansion of the urban economy (which seems unlikely in the Andes at present) and on natural resource subsidies from further afield—what we might call the ecological footprints of the development of the eastern United States (IIED 1995).

2. This work continues in a research, seminar, and information exchange program linking GIA (Chile), IIED (U.K.), ODI (U.K.), DESCO (Peru), CINEP (Colombia), UNITAS (Bolivia), and Fundagro (Ecuador).

3. I am quite aware that the following short paragraphs are less than a caricature, but I think they point to general patterns in these bodies of writing.

4. See Bebbington et al. 1992 and 1993a for Ecuador; Bebbington et al. 1995 for Bolivia; Trujillo 1993 and Tendler, Healey, and O'Laughlin 1988 for El Ceibo.

5. It should be noted that these stories of resource accessing are always slightly more convoluted than first meets the eye. This organization was able to access financial resources because its leader was a literacy trainer, taught by and acquainted with the chief of a literacy program in the area that had already worked with the same financier. This of course helped establish the contact between the organization and the funding agency. This is not an uncommon pattern.

6. It is interesting that the most modernizing and market-oriented of Latin

America's government extension programs (INDAP in Chile) has similarly begun to emphasize work with farmer organizations, conceiving their role in exactly the same way (see INDAP 1995).

7. Increasingly, however, donors are demanding more evidence that costs are being recovered and the value of capital maintained, partly because the donors are seeing their own funds dwindling. They have begun to realize they are not going to be able to continue subsidizing these organizations.

5. Organizing Experimenting Farmers

1. In practice, training was required as much to reinforce the skills of trainee agronomists to design and analyze on-farm trials as to teach these skills to the farmers.

2. This places weightier requirements on the paraprofessionals' skills, however, particularly regarding the training linked to reinforcing the CIALs' management capacities. At present, the most effective paraprofessionals in this respect are two young farmers with prior experience in community committees. These farmers have the self-confidence to speak in front of a group and have taught in the rural schools, although they have no training as teachers. As for their formal education, they have completed secondary school.

3. All references to dollars are to U.S. dollars.

4. Janssen et al. (1991:195–211) provide some data on the impact of on-farm research with beans carried out in Cajamarca, Peru, from 1982 to 1989. The research carried out by on-farm researchers and extensionists involved 10 on-farm trials in 1982, 30 demonstration plots in 1983, an unspecified number of demonstrations and on-farm variety trials in seven regions in 1984, 363 interviews with farmers and 51 experiments in 1985, and 53 trials in 1986. Estimates of the impact of this on-farm research determined that about 70 percent of the total was induced by the research program, resulting in additional bean production worth between $130,000.00 and $265,000.00, from an additional five thousand hectares of beans.

5. The fifty-five CIALs were not distributed geographically across this area on the basis of the requirements of particular agroecological niches or client groups for location-specific adaptive testing. The project's primary objective was methodology development; the secondary goal was to assess how different institutional linkages affect farmers' capacity to perform participatory research in agriculture. The CIALs were located geographically to facilitate comparison between institutional settings. As a result, existing CIALs are prob-

ably more highly concentrated than is warranted by the degree of location-specific diversity in the region.

6. Graf, Voss, and Nyabyenda (1991:56–57), studying an on-farm research program in Rwanda, found that farmer participation in on-farm research and a system of group meetings reduced the costs of on-farm varietal testing. These were estimated at $27,225.00 between 1986 and 1990, for work with about seventy-nine farmers in four communities, later reduced to about forty farmers in two communities. The research area covered about 31,500 hectares and twenty thousand families. Our calculations, based on the costs reported only for experimentation in this study (of which 80 percent were covered by a national program together with CIAT and for which salaries of researchers were costed at the level of national program salaries), range from about sixty-three to thirty-four dollars per farmer per year if we consider only the forty to seventy-nine farmers reported as actually involved in the trials. Since this study, the testing of climbing beans has diffused dramatically, so that the per capita costs of the on-farm research based on the population covered must have gone down significantly (Sperling et al. 1994).

7. For these materials, see Cartilla Nos. 1–12 and the video under IPRA, CIAT, in the bibliography to this volume.

6. Technologies for Sustainable Forest Management

1. All references to dollars are to U.S. dollars.

2. Design of a management plan requires consideration of biological and silvicultural factors, such as species composition and diameter distribution as well as site conditions, accessibility, and availability of logging equipment. All these factors interact to determine the optimal harvesting intensity for commercial timber and the maintenance of sufficient natural regeneration for future harvests. Hence, the management plan is not viewed from the narrow objective of providing a continuous supply of timber but explicitly considers the sustainability of the forest ecosystem and, thus, the conservation of wildlife habitats.

3. These include the use of improved cable logging systems, hand tools, and skidders to replace crowled tractors. Efficiency in handling is improved with equipment selected to match particular conditions. The area disturbed under traditional harvesting practices has been estimated to be as high as 50 percent; this has been reduced to 14 percent with the introduction of new technologies.

7. Indigenous Knowledge for Agricultural Development

1. Dr. Cernea recently wrote a new preface for the French and Spanish versions of the paper.

2. It is available through CIKARD and LEAD on the Internet. Back issues of the *Monitor* are now available through CIRAN's web site (http://www.nufficcs.nl/ciran/ikdm).

3. Available at http://www.iitap.iastate.edu/cikard/cikard.html. Another web site is available through CIESIN at http://www.ciesin.org/TG/thematic-home.html.

8. Local Knowledge Systems in Latin America

1. In this chapter, the term "local knowledge systems" (LKS) is used as an equivalent to "indigenous knowledge systems."

2. It is important to note that one limitation to such an overview is that not all of the projects related to local knowledge systems in Latin America have published results (such as many grassroots efforts) and that even the CIKARD documentation unit still lacks reports from some of those that have been published.

3. Unless otherwise noted, all references to dollars are to U.S. dollars.

Bibliography

Adams, W. M., and L. J. Slikkerveer, eds. 1997. *Indigenous Knowledge and Change in African Agriculture.* Studies in Technology and Social Change No. 26. Ames: CIKARD, Iowa State University.

Adriaanse, A. 1993. *Environmental Policy Performance Indicators.* The Netherlands: Sdu Publishers.

Ahmed, M. M. M., ed. 1994. *Indigenous Farming Systems, Knowledge and Practices in the Sudan.* Khartoum: Institute of African and Asian Studies, University of Khartoum.

Alcorn, J. B. 1984. "Development Policy, Forest, and Peasant Farms: Reflections on Huastec-managed Forests' Contributions to Commercial Production and Resource Conservation." *Economic Botany* 38(4):389–406.

Alexandratos, N., ed. 1988. *World Agriculture Toward 2000: A FAO Study.* Rome and London: FAO and Belhaven Press.

Altieri, M. A. 1987. "The Significance of Diversity in the Maintenance of the Sustainability of Traditional Agroecosystems." *ILEIA Newsletter* 3(2):3–7.

Altieri, M. A., M. K. Anderson, and L. C. Merrick. 1987. "Peasant Agriculture and the Conservation of Crop and Wild Plant Resources." *Conservation Biology* 1(1):49:58.

Altieri, M. A., and L. C. Merrick. 1987. "In Situ Conservation of Crop Genetic Resources Through Maintenance of Traditional Farming Systems." *Economic Botany* 41(1):86–96.

———. 1988. "Agroecology and In Situ Conservation of Native Crop Diversity in the Third World." In E. O. Wilson and F. M. Peter, eds., *Biodiversity.* Washington, D.C.: National Academy Press.

Altieri, M. A., and J. Trujillo. 1987. "The Agroecology of Corn Production in Tlaxcala, Mexico." *Human Ecology* 15(2):189–220.

Ames, A. I. 1995. "A Tutorial for the Electronic Matrix System: A System to Facilitate the Collection of Anthropological Data in the Field, Using a MacIntosh PowerBook 150 and Microsoft Office Software." Ames: CIKARD, Iowa State University.

Anderson, A. B. 1988. "Use and Management of Native Forests Dominated by Açaí Palm (*Euterpe oleracea* Mart.) in the Amazon Estuary." *Advances in Economic Botany* 6:144–54.

———. 1989. "Estratégias de uso da terra para reservas extrativistas da Amazônia." *Pará Desenvolvimento* 25:30–37.

———. 1990. "Extraction and Forest Management by Rural Inhabitants in the Amazon Estuary." In A. B. Anderson, ed., *Alternatives to Deforestation: Steps Toward Sustainable Use of the Amazon Rain Forest.* New York: Columbia University Press.

Anderson, A. B., and E. M. Ioris. 1992. "Valuing the Rain Forest: Economic Strategies by Small-Scale Forest Extractivists in the Amazon Estuary." *Human Ecology* 20(3):337–69.

Appleton, H., et al. 1993. "Women: Invisible Technologists." *Appropriate Technology* 20(2):1–5.

Ascher, W., and R. Healy. 1990. *Natural Resource Policymaking in Developing Countries: Environment, Economic Growth, and Income Distribution.* Durham, N.C.: Duke University Press.

Ashby, J. A. 1987. "The Effects of Different Types of Farmer Participation on the Management of On-Farm Trials." *Agricultural Administration and Extension Newsletter* 24:234–52.

———. 1990. *Evaluating Technology with Farmers: A Handbook.* CIAT, Palmira. Also in Spanish, Portuguese, and French.

Ashby, J. A., T. Gracia, M. Guerrero, C. A. Quirós, J. Roa, and J. Beltrán. 1995. "Organizing Experimenting Farmers for Participation in Agricultural Research and Technology Development." Paper presented at the Workshop on Traditional and Modern Approaches to Natural Resource Management in Latin America and the Caribbean, 25–26 April 1995, World Bank, Washington, D.C.

Ashby, J. A., and D. Pacheco. 1988. *Farmer Evaluations of Technology: A Handbook.* Colombia: IPRA Project, CIAT.

Ashby, J. A., C. A. Quirós, T. Gracia, M. Guerrero, and J. Roa. 1990. "Farmer Participation Early in the Evaluation of Agricultural Technologies." Paper presented at the Seminar on Reviving Local Self-Reliance: Challenges for Rural/Regional Development in Eastern and Southern Africa, 21–24 February 1990, Arusha, Tanzania.

Ashby, J. A., C. A. Quirós, and Y. M. Rivera. 1987. "Farmer Participation in On-Farm Varietal Trials." Agricultural Administration Research and Extension Network Paper No. 22. London: ODI.

Ashby, J. A., and L. Sperling. 1994. "Institutionalizing Participatory Client-Driven Research and Technology Development in Agriculture." Agricultural Administration Research and Extension Network Paper No. 49. London: ODI.

Awa, N. E. 1989. "Underutilization of Women's IK in Agricultural and Rural Development Programs: The Effect of Stereotypes." In D. M. Warren, L. J. Slikkerveer, and S. Oguntunji Titilola, eds., *Indigenous Knowledge Systems: Implications for Agriculture and International Development.* Studies in Technology and Social Change, No. 11. Ames: Technology and Social Change Program, Iowa State University.

Axinn, G. 1994. "Responses to Newsletter 29 in Agricultural Research and Extension Network." *Agricultural Administration and Extension Newsletter* 30:18.

Badri, B., and A. Badri. 1994. "Women and Biodiversity." *Development: Journal of the Society for International Development* 1:67–71.

Bahri, S. 1992. *L'agroforesterie, une alternative pour le développement de la plaine alluviale de l'Amazone: L'exemple de l'île de Careiro.* Doctoral dissertation, Université de Montpellier.

Barborak, J., and G. Green. 1987. "Implementing the World Conservation Strategy: Success Stories from Central America and Colombia." In D. D. Southgate and J. F. Disinger, eds., *Sustainable Resource Development in the Third World.* Boulder, Colo.: Westview Press.

Barreiro, J., ed. 1992. "Indigenous Economics: Toward a Natural World Order." *Akwe:kon Journal* (special issue) 9(2).

Bebbington, A. J. 1993. "Modernization from Below: An Alternative Indigenous Development?" *Economic Geography* 69(3):274–92.

———. 1995. "Organizing for Change—Organizing for Modernization? Campesino Federations and Technical Change in Andean and Amazonian Resource Management." Paper presented at the Workshop on Traditional and Modern Approaches to Natural Resource Management in Latin America and the Caribbean, 25–26 April 1995, World Bank, Washington, D.C.

———. 1996. "Organizations and Intensifications: Campesino Federations, Rural Livelihoods and Agricultural Technology in the Andes and Amazonia." *World Development* 24(7):1161–77.

Bebbington, A. J., H. Carrasco, L. Peralvó, G. Ramón, V. H. Torres, and J. Trujillo. 1992. "From Protest to Productivity: The Evolution of Indigenous Federations in Ecuador." *Grassroots Development* 16(2):11–21.

———. 1993a. "Rural Peoples' Knowledge, Farmers Organizations and Regional Development: Implications for Agricultural Research and Extension." Agricultural Administration Research and Extension Network Paper No. 41. London: ODI.

Bebbington, A. J., T. Domingo, A. Kopp, and J. Quisbert. 1995. *Campesino Federations, Food Systems and Rural Politics in Bolivia*. London: IIED.

Bebbington, A. J., D. Merrill-Sands, and J. Farrington. 1994. "Farmer and Community Organizations in Agricultural Research and Extension: Functions, Impacts and Questions." Agricultural Administration Research and Extension Network Paper No. 47. London: ODI.

Bebbington, A. J., G. Thiele, P. Davies, M. Prager, and H. Riveros, eds. 1993b. *Non-Governmental Organizations and the State in Latin America*. London: Routledge.

Bellon, M. R., and S. B. Brush. 1994. "Keepers of Maize in Chiapas, Mexico." *Economic Botany* 48(2):196–209.

Bentley, J. W. 1989. "What Farmers Don't Know Can't Help Them: The Strengths and Weaknesses of Indigenous Technical Knowledge in Honduras." *Agriculture and Human Values* 6(3):25–31.

Bentley, J. W., and K. L. Andrews. 1991. "Pests, Peasants, and Publications: Anthropological and Entomological Views of an Integrated Pest Management Program for Small-Scale Honduran Farmers." *Human Organization* 50(2):113–24.

Benzing, A. 1989. "Andean Potato Peasants Are 'Seed Bankers'." *ILEIA Newsletter* 5(4):12–13.

Berg, T. 1993. "The Science of Plant Breeding: Support or Alternative to Traditional Practices." In W. de Boef, K. Amanor, K. Wellard, and A. Bebbington, eds., *Cultivating Knowledge: Genetic Diversity, Farmer Experimentation and Crop Research*. London: ITP.

Berkes, F., and C. Folke. 1992. "A Systems Perspective on the Interrelations Between Natural, Human-Made and Cultural Capital." *Ecological Economics* 5:1–8.

———. 1994. "Linking Social and Ecological Systems for Resilience and Sustainability." Paper presented at the Workshop on Property Rights and the Performance of Natural Resource Systems, at the Badger International Institute of Ecological Economics, Royal Swedish Academy of Sciences, Stockholm, Sweden.

Berlin, B. 1992. *Ethnobiological Classification: Principles of Categorization of Plants and Animals in Traditional Societies*. Princeton, N.J.: Princeton University Press.

Bheenick, R., E. G. Bonkoungou, and others. 1989. *Successful Development in Africa: Case Studies of Projects, Programs, and Policies*. EDI Development Policy Case Series, Analytical Case Study No. 1. Washington, D.C.: World Bank.

Bibliography

Bidegaray, P., and R. E. Rhoades. N.d. *Los agricultores de Yurimaguas: Uso de la tierra y estrategias de cultivo en la selva peruana.* Documento No. 10. Lima, Peru: CIP y Centro de Investigación y Promoción Amazónica.

Biggs, S. 1988. *Resource-Poor Farmer Participation in Research: A Synthesis from Nine National Agricultural Research Programs.* The Hague: International Service for National Agricultural Research.

―――. 1989. "A Multiple Source of Innovation Model of Agricultural Research and Technology Promotion." Agricultural Administration Research and Extension Network Paper No. 6. London: ODI.

Biggs, S., and E. Clay. 1981. "Sources of Innovation in Agricultural Technology." *World Development* 9(4):321–36.

Bilsborrow, R. 1992. "Population Growth, Internal Migration, and Environmental Degradation in Rural Areas of Developing Countries." *European Journal of Population* 8:125–48.

Bilsborrow, R., and H. Ogendo. 1992. "Population-Driven Change and Land Use in Developing Countries," *Ambio* 21(1).

Bilsborrow, R., and M. Geores. 1992. *Rural Population Dynamics and Agricultural Development: Issues and Consequences Observed in Latin America.* Ithaca, N.Y.: CIIFAD.

Binswanger, H. 1987. *Fiscal and Legal Incentives with Environmental Effects on the Brazilian Amazon.* Washington, D.C.: Agricultural and Rural Development Department, World Bank.

―――. "Agricultural and Rural Development: Painful Lessons." Simon Brandt Address delivered at the thirty-second annual meeting of the Agricultural Economics Association of South Africa, 21 September 1994, Pretoria, South Africa.

Binswanger, H., K. Deininger, and G. Feder. 1995. "Power, Distortions, Revolt and Reform in Agricultural Land Relations." In J. Berhman and T. N. Srinivasan, eds., *Handbook of Development Economics.* Volume 3. Amsterdam: Elsevier.

Binswanger, H., and J. von Braun. 1991. "Technological Change and Commercialization in Agriculture: Impact on the Poor." *World Bank Research Observer* 6(1): 57–80.

Binswanger, H., and V. Ruttan. 1978. *Induced Innovation: Technology Institutions and Development.* Baltimore, Md.: Johns Hopkins University Press.

Biodiversity Support Program (BSP). 1995. *Geographic Biodiversity Conservation Investment Priorities in Latin America and the Caribbean.* Washington, D.C.: WWF.

Birkhaeuser, D., R. Evenson, and G. Feder. 1991. "The Economic Impact of Agricultural Extension: A Review." *Economic Development and Cultural Change* (April):607–50.

Blunt, P., and D. M. Warren, eds. 1996. *Indigenous Organizations and Development.* London: ITP.

Bojanic, A., M. E. Canedo, V. Gianotten, M. Morales, C. Ranaboldo, and W. Rijssenbeek. 1994. *Demandas campesinas: Manual para un analisis participativo.* La Paz, Bolivia: Embajada Real de los Paises Bajos.

Boserup, E. 1965. *The Conditions of Agricultural Growth.* Chicago, Ill.: Aldine.

————. 1981. *Population and Technology.* Oxford: Blackwell.

Brabo, M. J. C. 1979. "Palmiteiro de Muaná: Estudo sobre o processo de produção no beneficiamento do açaizeiro." *Boletim do Museu Paraense Emilio Goeldi* (nova série, antropologia) 73:1–29.

Brokensha, D., D. M. Warren, and O. Werner. 1980. *Indigenous Knowledge Systems and Development.* Lanham, Md.: University Press of America.

Brooke, L. F. 1993. *The Participation of Indigenous Peoples and the Application of Their Environmental and Ecological Knowledge in the Arctic Environmental Protection Strategy.* Ottawa: Inuit Circumpolar Conference.

Brown, L. 1989. *State of the World.* New York: Norton.

Brush, S. B. 1980. "Potato Taxonomies in Andean Agriculture." In D. Brokensha, D. M. Warren, and O. Werner, eds., *Indigenous Knowledge Systems and Development.* Washington, D.C.: University Press of America.

————. 1991. "A Farmer-Based Approach to Conserving Crop Germplasm." *Economic Botany* 45(2):153–65.

Brush, S. B., and D. Stabinsky, eds. 1996. *Valuing Local Knowledge: Indigenous People and Intellectual Property Rights.* Washington, D.C.: Island Press.

Bunch, R. 1985. *Two Ears of Corn: A Guide to People-Centered Agricultural Improvement.* Oklahoma City, Okla.: World Neighbors.

Bunders, J., B. Haverkort, and W. Hiemstra, eds. 1996. *Biotechnology: Building on Farmers' Knowledge.* London: Macmillan.

Bunyard, P. 1994. "Bringing Back the Balance: Alternative Economics for the Colombian Amazon." *ILEIA Newsletter* 10(2):10–11.

Busch, L., and W. Lacy. 1983. *Science, Agriculture, and the Politics of Research.* Boulder, Colo.: Westview Press.

Butler, L., and J. Waud. 1990. "Strengthening Extension Through the Concepts of Farming Systems Research and Extension (FSRE) and Sustainability." *Journal of Farming Systems Research-Extension* 1(1):77–92.

Carroll, T. 1992. *Intermediary NGOs: The Supporting Link in Grassroots Development.* West Hartford, Conn.: Kumarian Press.

Cashman, K. 1989. "Agricultural Research Centers and Indigenous Knowledge Systems in a Worldwide Perspective: Where Do We Go from Here?" In D. M. Warren, L. J. Slikkerveer, and S. Oguntunji Titilola, eds., *Indigenous Knowledge Systems: Implications for Agriculture and International Development.* Studies in Technology and Social Change Program No. 11. Ames: Iowa State University Research Foundation.

Castro, P. 1995. *Facing Kirinyaga.* London: ITP.

CEPAL (Comisión Económica para América Latina). 1983. *Expansión de la frontera agropecuaria y medio ambiente en América Latina.* Santiago: CEPAL.

———. 1988. *Restrictions on Sustained Development in Latin America and the Caribbean and the Requisites for Overcoming Them.* Santiago: CEPAL LC/G. 1988/Rev.

———. 1989. *Latin America and the Caribbean: The Management of Water Scarcity.* Santiago: CEPAL LC/R.774.

———. 1990. *Magnitud de la pobreza en América Latina en los años ochenta.* Santiago: CEPAL.

———. 1995. *Panorama social de América Latina.* Santiago: CEPAL.

Cernea, M. 1991. *Putting People First: Sociological Variables in Rural Development.* 2d edition. New York: Oxford University Press.

———. 1992. *The Building Blocks of Participation: Testing Bottom-up Planning.* World Bank Discussion Paper No. 166. Washington, D.C.: World Bank.

Cernea, M., J. Coulter, and J. Russell. 1985. "Building the Research-Extension-Farmer Continuum: Some Current Issues." In M. Cernea, J. Coulter, and J. Russell, eds., *Research-Extension-Farmer: A Two-Way Continuum for Agricultural Development.* Washington, D.C.: World Bank.

CGIAR (Consultative Group on International Agricultural Research). 1985. *Summary of IARCs: A Study of Achievements and Potential.* Washington, D.C.: CGIAR.

———. 1990. *A Possible Expansion of CGIAR.* Washington, D.C.: Technical Advisory Committee.

———. 1993. "Indigenous Knowledge." *People and Plants: The Development Agenda.* Rome: CGIAR.

Chambers, R. 1983. *Rural Development: Putting the Last First.* London: Longman.

———. 1994. "Foreword." In I. Scoones and J. Thompson, eds., *Beyond Farmer First.* London: ITP.

————. 1997. *Whose Reality Counts? Putting the First Last.* London: ITP.

Chambers, R., ed. 1979. "Rural Development: Whose Knowledge Counts?" *IDS Bulletin* (special issue) 10(2).

Chambers, R., A. Pacey, and L. Thrupp, eds. 1989. *Farmer First: Farmer Innovation and Agricultural Research.* London: ITP.

Chambers, R., and J. Pretty. 1994. "Are the International Agricultural Research Centers Tackling the Crucial Issues of Poverty and Sustainability?" *International Agricultural Development* (November–December).

Chang, J. H. 1977. "Tropical Agriculture: Crop Diversity and Crop Yields." *Economic Geography* 53:241–54.

Chapin, M. 1988. "The Seduction of Models: Chinampa Agriculture in Mexico." *Grassroots Development* 12(1):8–17.

CIAD (Center for Integrated Agricultural Development). 1994. *Indigenous Knowledge Systems and Rural Development in China: Proceedings of the Workshop.* Beijing: Beijing Agricultural University.

CIMMYT (International Maize and Wheat Improvement Center). 1988. *Annual Report.* Mexico City: CIMMYT.

Clawson, D. L. 1985. "Harvest Security and Intraspecific Diversity in Traditional Tropical Agriculture." *Economic Botany* 39(1):56–67.

Clay, J. W., and C. R. Clement. 1993. *Selected Species and Strategies to Enhance Income Generation from Amazonian Forests.* FO:Misc/93/6. Working Paper. Rome: FAO.

Cochrane, W. 1979. *The Development of American Agriculture.* Minneapolis: University of Minnesota Press.

Collinson, M. 1985. "Farming Systems Research: Diagnosing the Problems." In M. Cernea, J. Coulter, and J. Russell, eds., *Research-Extension-Farmer: A Two-Way Continuum for Agricultural Development.* Washington, D.C.: World Bank.

————. "On Farm Research and Agricultural Research and Extension Institutions." Agricultural Administration Research and Extension Network Paper No. 17. London: ODI.

Collion, M. H. 1995. "On Building a Partnership in Mali Between Farmers and Researchers." Agricultural Administration Research and Extension Network Paper No. 54. London: ODI.

Commoner, B. 1971. *The Closing Circle: Nature, Man and Technology.* New York: Knopf.

Conway, G., and J. Pretty. 1991. *Unwelcome Harvest: Agriculture and Pollution.* London: Earthscan.

Bibliography

Cook, C., and M. Grut. 1989. *Agroforestry in Sub-Saharan Africa: A Farmers' Perspective*. World Bank Technical Paper No. 112. Washington, D.C.: World Bank.

Cortes, J. 1995. "Seed Systems Development in Peru." Paper presented at the Workshop on Traditional and Modern Approaches to Natural Resource Management in Latin America and the Caribbean, 25–26 April 1995, World Bank, Washington, D.C.

COSEFORMA (Proyecto de Cooperación en los Sectores Forestal y Maderero). 1993. *Día de campo. Bosque secundario. Finca Miguel Carazo. Pejibaye de San Rafael de Guatuso*. Informe de campo. Costa Rica.

————. 1994. *Inventario forestal de la Región Huetar Norte. Resumen de resultados. Basado en el informe de consultoría de la GWB*. Costa Rica.

Costanza, R., and H. Daly. 1992. "Natural Capital and Sustainable Development." *Conservation Biology* 6(1):37–46.

Crosson, P. 1983. *Soil Erosion in Developing Countries: Amounts, Consequences and Policies*. Madison: University of Wisconsin, Center for Resource Policy Studies.

Cultural Survival Quarterly. 1995. *Geomatics: Who Needs It? Cultural Survival Quarterly* 18(4).

Cunningham, A. B. 1991. "Indigenous Knowledge and Biodiversity: Global Commons or Regional Heritage?" *Cultural Survival Quarterly* 15(3):4–8.

Davis, S. H., ed. 1993. *Indigenous Views of Land and the Environment*. World Bank Discussion Paper No. 188. Washington, D.C.: World Bank.

de Boef, W. S., K. Amanor, K. Wellard, and A. Bebbington, eds. 1993. *Cultivating Knowledge: Genetic Diversity, Farmer Experimentation and Crop Research*. London: ITP.

De Janvry, A. 1981. *The Agrarian Question and Reformism in Latin America*. Baltimore, Md.: Johns Hopkins University Press.

del Valle, J. 1804. *Instrucción sobre la plaga de langosta: Medios de exterminala, o de disminuir sus efectos, y de precaber la escasez de comestibles*. Gobierno Superior de Guatemala: D. Ignacio Beteta.

DenBiggelaar, C. 1991. "Farming Systems Development: Synthesizing Indigenous and Scientific Knowledge Systems." *Agriculture and Human Values* 8(1):25–36.

Denevan, W. M. 1966. *The Aboriginal Cultural Geography of the Llanos de Mojos of Bolivia*. Berkeley and Los Angeles: University of California Press.

————. 1970. "The Aboriginal Population of Western Amazonia in Relation to Habitat and Subsistence." *Revista Geográfica* 72:61–86.

Bibliography

Denning, G. 1985. "Integrating Agricultural Extension Programs with Framing Systems Research." In M. Cernea, J. Coulter, and J. Russell, eds., *Research-Extension-Farmer: A Two-Way Continuum for Agricultural Development.* Washington, D.C.: World Bank.

DeWalt, B. R. 1978. "Appropriate Technology in Rural Mexico: Antecedents and Consequences of an Indigenous Peasant Innovation." *Technology and Culture* 19:32–52.

———. 1979. *Modernization in a Mexican Ejido: A Study in Economic Adaptation.* Cambridge and New York: Cambridge University Press.

———. 1984. "International Development Paths and Policies: The Cultural Ecology of Development." *Rural Sociologist* 4:255–68.

———. 1988. "The Cultural Ecology of Development: Ten Precepts for Survival." *Agriculture and Human Values* 5(1–2):112–23.

———. 1989. "Halfway There: Social Science in Agricultural Development and the Social Science of Agricultural Development." In C. McCorkle, ed., *Social Sciences in International Agricultural Research: Lessons from the CRSPs.* Boulder, Colo.: Lynne Rienner.

———. 1991a. "The Cultural Ecologist Concept of Justice." In P. B. Thompson and B. A. Stout, eds., *Beyond the Large Farm: Ethics and Research Goals for Agriculture.* Boulder, Colo.: Westview Press.

———. 1991b. "Anthropology, Evolution, and Agricultural Development." In G. C. Johnson and J. T. Bonnen, with D. Fienup, C. L. Quance, and N. Schaller, eds., *Social Science Agricultural Agendas and Strategies.* East Lansing: Michigan State University Press.

———. 1994. "Using Indigenous Knowledge to Improve Agriculture and Natural Resource Management." *Human Organization* 53(2):123–31.

———. 1995. "Using Indigenous Knowledge to Improve Agriculture and Natural Resource Management." Paper presented at the Workshop on Traditional and Modern Approaches to Natural Resource Management in Latin America and the Caribbean, 25–26 April 1995, World Bank, Washington, D.C.

DeWalt, B. R., K. M. DeWalt, with E. Adelski, S. Duda, M. Fordham, and K. Thompson. 1982. *Cropping Systems in Pespire, Southern Honduras.* Farming Systems Research in Southern Honduras, Report No. 1, University of Kentucky Departments of Anthropology and Rural Sociology. Lexington, Ky.: INTSORMIL Project.

DeWalt, B. R., P. Vergne, and M. Hardin. 1993. "Population, Aquaculture and Environmental Deterioration: The Gulf of Fonseca, Honduras." Paper prepared for the René Dubois Center Forum on Population, Environment and Development, 22–23 September 1993, New York Academy of Medicine.

DeWalt, K. M., and B. R. DeWalt. 1989. "Social Science Approaches to Including Nutrition Research in the Sorghum/Millet CRSP." In C. McCorkle, ed., *Social Sciences in International Agricultural Research: Lessons from the CRSPs.* Boulder, Colo.: Lynne Rienner.

Diamond, J. 1987. "The Worst Mistake in the History of the Human Race." *Discover* (May):64–66.

Dinerstein, E., D. M. Olson, D. J. Graham, A. L. Webster, S. A. Primm, M. P. Bookbinder, and G. Ledec. 1995. *A Conservation Assessment of the Terrestrial Ecoregions of Latin America and the Caribbean.* Washington, D.C.: World Bank. Published in association with the World Wildlife Fund.

Dommen, A. 1988. *Innovation in African Agriculture.* Boulder, Colo.: Westview Press.

———. 1989. "A Rationale for African Low-Resource Agriculture in Terms of Economic Theory." In D. M. Warren, L. J. Slikkerveer, and S. Oguntunji Titilola, eds., *Indigenous Knowledge Systems: Implications for Agriculture and International Development.* Studies in Technology and Social Change Program No. 11. Ames: Iowa State University Research Foundation.

Dover, M., and L. Talbot. 1987. *To Feed the Earth: Agroecology for Sustainable Development.* Washington, D.C.: WRI.

Dowdeswell, E. 1993. "Walking in Two Worlds." Address presented at the InterAmerican Indigenous People's Conference, 18 September 1993, Vancouver, Canada.

Drinkwater, M. 1994a. "Developing Interaction and Understanding: RRA and Farmer Research Groups in Zambia." In I. Scoones and J. Thompson, eds., *Beyond Farmer First: Rural People's Knowledge, Agricultural Research and Extension Practice.* London: ITP and IIED.

———. 1994b. "Knowledge, Consciousness and Prejudice." In I. Scoones and J. Thompson, eds., *Beyond Farmer First: Rural People's Knowledge, Agricultural Research and Extension Practice.* London: ITP and IIED.

Dugue, P. 1993. "The Senegalese Institute for Agricultural Research (ISRA) and the Fatick Region Farmers' Association." In K. Wellard and J. Copestake, eds., *NGOs and the State in Africa.* London: Routledge.

Duncan, A. 1993. "Financing Agricultural Services in Sub-Saharan Africa." *Food Policy* 18(6):433–65.

Eckholm, E. P. 1976. *Losing Ground: Environmental Stress and World Food Prospects.* New York: Norton.

Bibliography

EPA (Environmental Protection Agency). 1994. *A Conceptual Framework to Support the Development and Use of Environmental Information.* EPA 230-R-94-012. Washington, D.C.: EPA.

Erickson, C. L. 1994. "Raised Fields as a Sustainable Agricultural System from Amazonia." Paper presented in the Symposium on Recovery of Indigenous Technology and Resources in Bolivia, Eighteenth International Congress of the Latin American Studies Association, Atlanta, Georgia.

Esman, M., and N. Uphoff. 1984. *Local Organizations: Intermediaries in Rural Development.* Ithaca, N.Y.: Cornell University Press.

Eyzaguirre, P., and M. Iwanaga, eds. 1996. *Participatory Plant Breeding.* Proceedings of the Workshop on Participatory Plant Breeding, 26–29 July 1995, Wageningen, the Netherlands. Rome: International Plant Genetic Resources Institute.

Fairhead, J. 1990. "Fields of Struggle: Towards a Social History of Farming Knowledge and Practice in a Bwisha Community, Kivu, Zaire." Ph.D. dissertation, School of African Studies, University of London.

FAO (Food and Agriculture Organization of the United Nations). 1985. *Intensive Multiple-Use Forest Management in the Tropics: Analysis of Case Studies from India, Africa, Latin America and the Caribbean.* FAO: Forestry Paper No. 55.

———. 1986a. *Food and Fruit-Bearing Forest Species. 3: Examples from Latin America.* FAO: Forestry Paper No. 44-3.

———. 1986b. *Some Medicinal Forest Plants of Africa and Latin America.* FAO: Forestry Paper No. 67.

———. 1987. *Small-Scale Forest-Based Processing Enterprises.* FAO: Forestry Paper No. 79.

———. 1988a. *Potentials for Agriculture and Rural Development in Latin America and the Caribbean.* Rome: FAO.

———. 1988b. *Fruit-Bearing Forest Trees.* FAO: Forestry Paper No. 34.

———. 1989. *Small-Scale Harvesting Operations of Wood and Non-Wood Forest Products Involving Rural People.* FAO: Forestry Paper No. 87.

———. 1990. *Malnutrition in the Latin American and Caribbean Region: Causes and Prevention.* LARC 90/4. Santiago: FAO.

———. 1991a. *Sustainable Agriculture and Rural Development in Latin America and the Caribbean.* Regional Document No. 3. Rome: FAO.

———. 1991b. *Non-Wood Forest Products: The Way Ahead.* FAO: Forestry Paper No. 97.

———. 1992. *Agrostat.* Rome: FAO.

———. 1993. *Common Forest Resource Management.* Rome: FAO.

————. 1994. "Appendix 7: Forest Genetic Resources Priorities." *Report on the 8th Session of FAO Panel of Experts on Forest Gene Resources.* Rome: FAO.

Farrington, J., and A. J. Bebbington. 1994. *From Research to Innovation: Getting the Most from Interaction with NGOs in Farming Systems Research and Extension.* Gatekeeper Series Paper No. 43. London: IIED.

Farrington, J., A. J. Bebbington, K. Wellard, and D. J. Lewis. 1993. *Reluctant Partners? Non-Governmental Organizations: The State and Sustainable Agricultural Development.* London: Routledge.

Farrington, J., and A. Martin. 1988. "Farmer Participation in Agricultural Research: A Review of Concepts and Practices." Agricultural Administration Unit Occasional Paper No. 9. London: ODI.

Fearnside, P. 1986. *Human Carrying Capacity of the Brazilian Rainforest.* New York: Columbia University Press.

Fearnside, P. M., and J. M. Rankin. 1982. "The New Jarí: Risks and Prospects of a Major Amazonian Development." *Interciencia* 7:329–39.

Fedlmeir, C. 1995. "La importancia del bosque secundario en la Zona Norte de Costa Rica." Summary. Unpublished report, Ciudad Quesada, Costa Rica.

Felber, R., and C. Foletti. 1989. "Estudio sobre agricultura migratoria en la zona de Guajiquiro, Opatoro." Programa Marcala-Goascoran, Secretaría de Recursos Naturales y Cooperación Suiza del Desarrollo (COSUDE), Marcala, Honduras.

Fernández, M., ed. 1989. *El trabajo familiar y el rol de la mujer en la ganadería en comunidades alto-andinas de producción mixta.* Serie Comunidades. Reporte Técnico No. 101. Lima, Perú: Lluvia Editores. Proyecto de Validación de Tecnología en Comunidades, Huancayo.

Fernández-Baca, E. 1994. "The Role of Women in Livestock Production in the Mantaro Valley (Perú)." *Indigenous Knowledge and Development Monitor* (special issue) 2(3):20.

Feyerabend, P. 1975. *Against Method.* Atlantic Highlands, N.J.: Humanities Press.

Finnegan. B., C. Sabogal, C. Reiche, and I. Hutchinson. 1993. "Los bosques húmedos tropicales de América Central: Su manejo sostenible es posible y rentable." *Revista Forestal Centroamericana,* Costa Rica (edición especial) 6(2):17–27.

Flora, C. B. 1992. "Reconstructing Agriculture: The Case for Local Knowledge." *Rural Sociology* 57:92–97.

Forest Stewardship Council. 1994. "Principles and Criteria for Natural Forest Management." Board approved version, June 1994, Oaxaca, México. Unpublished document.

Fortmann, L. 1989. "Women's Role in Small Farm Agriculture." In M. A. Altieri and S. B. Hecht, eds., *Agroecology and Small Farm Development.* Boston: CRC Press.

Fortmann, L., and D. Rocheleau. 1985. "Women and Agroforestry: Four Myths and Three Case Studies." *Agroforestry Systems* (2):253–72.

Fox, J. 1992. "Democratic Rural Development: Leadership Accountability in Regional Peasant Organizations." *Development and Change* 23(2):1–36.

Franke, R., and B. Chasin. 1980. *Seeds of Famine: Ecological Destruction and the Development Dilemma in the West African Sahel.* Montclair, N.J.: Allanheld, Osmun and Company.

Frechione. J., D. A. Posey, and L. Francelino da Silva. 1989. "The Perception of Ecological Zones and Natural Resources in the Brazilian Amazon: An Ethnoecology of Lake Coari." In D. A. Posey and W. Balée, eds., *Resource Management in Amazonia: Indigenous and Folk Strategies (Advances in Economic Botany).* Vol. 7. Bronx, N.Y.: New York Botanical Garden.

Friedmann, H. 1994. "Distance and Durability: Shaky Foundations of the World Food Economy." In P. McMichael, ed., *The Global Restructuring of Agro-Food Systems.* Ithaca, N.Y.: Cornell University Press.

Fujisaka, S. 1991. *Seeds of Famine: Ecological Destruction and the Development Dilemma in the West African Sahel.* Montclair, N.J.: Allanheld, Osmun and Co.

———. 1995. "Incorporating Farmers' Knowledge in International Rice Research." In D. M. Warren, L. J. Slikkerveer, and D. Brokensha, eds., *The Cultural Dimension of Development: Indigenous Knowledge Systems.* London: ITP.

Gadgil, M. 1993. "Biodiversity and India's Degraded Lands." *Ambio* 22(2–3): 167–72.

Gallopín, G. C. 1994. *Agroecosystem Health: A Guiding Concept for Agricultural Research?* Proceedings of the International Workshop in Agroecosystem Health, Guelph, Ontario. Published with the support of the International Development Research Center through the University of Guelph Agroecosystem Health Project.

Gallopín, G. C., P. Gutman, and H. Maletta. 1988. "Global Impoverishment, Sustainable Development and the Environment." Report to the International Development Research Center Project: Global Impoverishment and Sustainable Development, Ecological Systems Analysis Group, S. C. Bariloche, Rio Negro, Argentina.

Garrett, J. 1995. "A 2020 Vision for Food, Agriculture, and the Environment in Latin America." International Food Policy Research Institute Discussion Paper No. 6. Washington, D.C.: IFPRI.

Gianotten, V., and W. Rijssenbeek, eds. 1995. *Peasant Demands: Manual for Participatory Analysis.* The Hague: Ministry of Foreign Affairs.

Gliessman, S., R. Garcia, and M. Amador. 1981. "The Ecological Basis for the Application of Traditional Agricultural Technology in the Management of Tropical Agro-Ecosystems." *Agro-Ecosystems* 7:173–75. Amsterdam: Elsevier Scientific Publishing Company.

Goldrich, D., and D. Carruthers. 1992. "Sustainable Development in Mexico? The International Politics of Crisis and Opportunity." *Latin American Perspectives* 72(19):97–122.

Goulding, M., N. J. H. Smith, and D. Mahar. 1996. *Floods of Fortune: Ecology and Economy Along the Amazon.* New York: Columbia University Press.

Graf, W., J. Voss, and P. Nyabyenda. 1991. "Climbing Bean Introduction in Southern Rwanda." In R. Tripp, ed., *Planned Change in Farming Systems: Progress in On-Farm Research.* Chichester, England: John Wiley and Sons.

Grimshaw, R. G. 1993. "The Vetiver Network." *Development: Journal of the Society for International Development* 3:64.

Grit, M. P. 1993. "Everyone's Knowledge Counts: Learning About Trees in Rio Grande do Norte." *Forests, Trees and People Newsletter* 19:10–13.

Groenfeldt, D. 1991. "Building on Tradition: Indigenous Irrigation Knowledge and Sustainable Development in Asia." *Agriculture and Human Values* 8(1): 114–20.

Guerrero, M. P. 1991. *The IPRA Method: Video Study Guide.* CIAT, Palmira. Also in Spanish.

Guerrero, M. P., J. Ashby, and T. Gracia. 1993. *Farmer Evaluation of Technology: Preference Ranking.* Instructional Unit No. 2. CIAT, Palmira. Also in Spanish.

Gupta, A. 1990. *Honey Bee* (quarterly journal devoted to indigenous knowledge). Ahmedabad, India: Indian Institute of Management.

———. 1993. "Honey Bee: An Accountable Global Network of Grassroots Innovators and Experimenters." *Development: Journal of the Society for International Development* 3:64–65.

Hammond, A., A. Adriaanse, E. Rodenburg, D. Bryant, and R. Woodward, eds. 1994. *Environmental Indicators: A Systematic Approach to Measuring and Reporting on Environmental Policy Performance in the Context of Sustainable Development.* Project on Indicators of Sustainable Development of the Scientific Committee on Problems of the Environment (SCOPE). Washington, D.C.: WRI.

Hansen, M. 1987. *Escape from the Pesticide Treadmill: Alternatives to Pesticides in Developing Countries.* Mount Vernon, N.Y.: Institute for Consumer Policy Research.

Hardon, J. J., and W. S. de Boef. 1993. "Linking Farmers and Breeders in Local Crop Development." In W. S. de Boef, K. Amanor, and K. Wellard, eds., *Cultivating Knowledge: Genetic Diversity, Farmer Experimentation and Crop Research*. London: ITP.

Haverkort, B., W. Hiemstra, C. Reijntjes, and S. Essers. 1988. "Strengthening Farmers' Capacity for Technology Development." *ILEIA Newsletter* 4(3):3–7.

Haverkort, B., and D. Millar. 1994. "Constructing Diversity: The Active Role of Rural People in Maintaining and Enhancing Biodiversity." *Etnoecolgica* 2–3:51–63.

Haverkort, B., J. van der Kamp, and A. Waters-Bayer, eds. 1991. *Joining Farmers' Experiments*. London: ITP.

Hayami, Y., and V. Ruttan. 1985. *Agricultural Development: An International Perspective*. Baltimore, Md.: Johns Hopkins University Press.

———. 1987. "Population Growth and Agricultural Productivity." In G. D. Johnson and R. D. Lee, eds., *Population Growth and Economic Development: Issues and Evidence*. Madison: University of Wisconsin Press.

Hazell, P., and C. Ramasamy. 1989. *The Green Revolution Reconsidered: The Impact of the High-Yielding Rice Varieties in South India*. Baltimore, Md.: Johns Hopkins University Press.

Hecht, S. B., and D. A. Posey. 1989. "Preliminary Results of Soil Management Techniques of the Kayapó Indians." *Advances in Economic Botany* 7:174–88.

Heinrich G., F. Worman, and C. Koketso. 1991. "Integrating FPR with Conventional On-Farm Research Programs: An Example from Botswana." *Journal for Farming Systems Research-Extension* 2:1–15.

Herbert, J. 1993. "A Mail-Order Catalog of Indigenous Knowledge." *Ceres: The FAO Review* 25(5):33–37.

Hess, C. 1997. *Hungry for Hope: On the Cultural and Communicative Dimensions of Development in Highland Ecuador*. London: ITP.

Hightower, J. 1973. *Hard Tomatoes, Hard Times*. Cambridge, Mass.: Schenkman.

Hiraoka, M. 1985. "Changing Floodplain Livelihood Patterns in the Peruvian Amazon." *Tsukuba Studies in Human Geography* 9:243–75.

Hochman, E., and D. Zilberman. 1986. "Optimal Strategies of Development Processes of Frontier Environments." *Science of the Total Environment* 55(1):111–20.

Horton, D. 1986. "Farming Systems Research: Twelve Lessons from the Mantaro Valley Project." *Agricultural Administration* (23):93–107.

Hoskins, M. 1984. "Observations on Indigenous and Modern Agro-Forestry

Bibliography

Activities in West Africa." In J. K. Jackson, ed., *Social, Economic, and Institutional Aspects of Agro-forestry.* Tokyo: United Nations University.

IDRC (International Development Research Center). 1993. "Indigenous and Traditional Knowledge." *IDRC Reports* (special issue) 21(1).

IIED (International Institute for Environment and Development). 1995. *PLA Notes: Notes on Participatory Learning and Action* No. 22. London: Sustainable Agriculture Program, IIED.

INDAP (Instituto de Desarrollo Agropecuario). 1995. *Plan de modernización: Programa de Transferencia Tecnológica (PTT): 1995–1997.* Santiago: INDAP.

Innis, D. 1997. *Intercropping: The Scientific Basis for Traditional Agriculture.* London: ITP.

IPGRI (International Plant Genetic Resources Institute). 1993. "Rural Development and Local Knowledge: The Case of Rice in Sierra Leone." *Geneflow.* Rome: IPGRI.

―――. 1994. Geneflow News: "Secret Garden of the Kayapó." *Geneflow.* Rome: IPGRI.

IPRA, CIAT. 1993a. *El Ensayo.* Cartillas para CIAL No. 1. English translation, *The Experiment.* Cali, Colombia: CIAT.

―――. 1993b. *Los comités de investigación agropecuaria local.* Cartillas para CIAL No. 2. English translation, *Local Agricultural Research Committees.* Cali, Colombia: CIAT.

―――. 1993c. *El diagnóstico.* Cartillas para CIAL No. 3. English translation, *The Diagnosis.* Cali, Colombia: CIAT.

―――. 1993d. *El objetivo del ensayo.* Cartillas para CIAL No. 4. English translation, *The Objective of the Experiment.* Cali, Colombia: CIAT.

―――. 1993e. *La planeación del ensayo.* Cartillas para CIAL No. 5. English translation, *Planning the Experiment.* Cali, Colombia: CIAT.

―――. 1993f. *La evaluación del ensayo.* Cartillas para CIAL No. 6. English translation, *Evaluating the Experiment.* Cali, Colombia: CIAT.

―――. 1993g. *Cosas que pueden pasar.* Cartillas para CIAL No. 7. English translation, *Things That Can Go Wrong.* Cali, Colombia: CIAT.

―――. 1993h. *Compartimos los resultados de nuestro ensayo.* Cartillas para CIAL No. 8. English translation, *Let's Share the Results of Our Experiment.* Cali, Colombia: CIAT.

―――. 1993i. *Un caso real.* Cartillas para CIAL No. 9. English translation, *A Real-Life Case Study.* Cali, Colombia: CIAT.

―――. 1993j. *Cómo manejar los gastos del ensayo* Cartillas para CIAL No. 10. English translation, *How to Manage Costs of the Experiment.* Cali, Colombia. CIAT.

———. 1993k. *Las experiencias también cuentan.* Cartillas para CIAL, No. 11. English translation, *We Learn From Our Experiences.* Cali, Colombia: CIAT.

———. 1993l. *Sabiendo a tiempo si vamos bien* Cartillas para CIAL, No. *12. Let's Find Out If We're Doing Things Right.* Cali, Colombia: CIAT.

———. 1993m. *Guias para conocer nuestro camino.* Cartillas para CIAL No.13. English translation, *Guidelines for Knowing Our Way.* Cali, Colombia: CIAT.

The IPRA Method. Video available in English and Spanish, from IPRA Project (Attn. T. Gracia), CIAT, Apartado Aereo 6713, Cali, Colombia, South America.

Iqbal, M. 1993. *International Trade in Non-Wood Forest Products: An Overview.* FAO Working paper FO:Misc-93-11.

Irvine, D. 1987. *Resource Management by the Runa Indians of the Ecuadorian Amazon.* Ph.D. dissertation, Stanford University.

ISD/RIVM/WRI (Institute for Sustainable Development/National Institute of Public Health and the Environment/World Resources Institute). 1995. Report of the Ecosystem Indicators Workshop, 16–18 March 1995. Institute for Sustainable Development, Menlo Park, California.

IUCN/UNEP/WWF (World Conservation Union/United Nations Environment Program/World Wildlife Fund). 1991. *Caring for the Earth: A Strategy for Sustainable Living.* Gland, Switzerland: IUCN.

Jaffee, S., and J. Srivastava. 1992. *Seed System Development: The Appropriate Roles of the Private and Public Sectors.* World Bank Discussion Paper No. 167. Washington, D.C.: World Bank.

Janssen W., N. Ruiz de Londoño, J. A. Beltrán, and J. Woolley. 1991. "On-Farm Research in Support of Varietal Diffusion: Bean Production in Cajamarca, Perú." In R. Tripp, ed., *Planned Change in Farming Systems: Progress in On-Farm Research.* Chichester, England: John Wiley and Sons.

Janzen, D. H. 1972. "The Uncertain Future of the Tropics." *Natural History* (November):80–89.

Jiggins, J. 1994. *Changing the Boundaries: Women-Centered Perspectives on Population and the Environment.* Washington, D.C.: Island Press.

Jiménez, L., and D. Hobbs. 1985. "Social Organization of Alpaca Production in Perú: A Case Study." Small Ruminant Collaborative Research Support Program, Technical Report Series No. 72. Department of Rural Sociology, University of Missouri–Columbia, Columbia, Missouri.

Jodha, N. 1990. "Mountain Agriculture: The Search for Sustainability." *Journal of Farming Systems Research-Extension* 1(1):55–75.

Johnson, A. 1983. "Machiguenga Gardens." In H. B. Hames and W. T. Vickers, eds., *Adaptive Responses of Native Amazonians*. New York and London: Academic Press.

Johnston, B. F., and W. C. Clark. 1982. *Redesigning Rural Development: A Strategic Perspective*. Baltimore, Md.: Johns Hopkins University Press.

Johnston, B. F., and P. Kilby. 1975. *Agriculture and Structural Transformation: Economic Strategies in Late-Developing Countries*. New York: Oxford University Press.

Jordan, C. 1985. *Nutrient Cycling in Tropical Forest Ecosystems*. Chichester, England: John Wiley and Sons.

Kainer, K. A., and M. L. Duryea. 1992. "Tapping Women's Knowledge: Plant Resource Use in Extractive Reserves, Acre, Brazil." *Economic Botany* 46(4):408–25.

Karr, J. R. 1995. "Ecological Integrity and Ecological Health Are Not the Same." In P. Schulze, ed., *Engineering with Ecological Constraints*. Washington, D.C.: National Academy Press.

Kevin, C., and G. Schreiber. 1992. *The Population, Agriculture, Environmental Nexus in Sub-Saharan Africa*. Agriculture and Rural Development Series No. 1. Washington, D.C.: World Bank, Africa Technical Department, Agriculture Division.

Kilahama, F. B. 1994. "Indigenous Ecological Knowledge: A Vital Tool for Rural Extension Strategies." *Forests, Trees and People Newsletter* (24):30–35.

Kirmse, R. D. 1993. *Prospects for Improved Management of Natural Forests in Latin America*. LATEN Dissemination Note No. 9. Washington, D.C.: World Bank, Latin America Technical Department, Environment Division.

Kishor N., and L. F. Constantino. 1993. *Forest Management and Competing Land Uses: An Economic Analysis for Costa Rica*. LATEN Dissemination Note No. 7. Washington, D.C.: World Bank, Latin America Technical Department, Environment Division.

Kloppenburg, J., Jr. 1991. "Social Theory and the De/Reconstruction of Agricultural Science: Local Knowledge for an Alternative Agriculture." *Rural Sociology* 56:519–48.

Kramer, S. N. 1956. *From the Tablets of Sumer: Twenty-Five Firsts in Man's Recorded History*. Indian Hills, Colo.: Falcon's Wing Press.

Kroma, S. 1995. "Popularizing Science Education in Developing Countries Through Indigenous Knowledge." *Indigenous Knowledge and Development Monitor* 3(3):13–15.

Krueger, A., M. Schiff, and A. Valdes. 1993. *The Political Economy of Agricultural Price Intervention in Latin America*. Washington, D.C.: World Bank.

Bibliography

Kuznets, S. 1980. "Driving Forces of Economic Growth: What Can We Learn from History?" *Weltwirtschaftliches Archiv* 16:409–31.

Lalonde, A. 1993. *Final Report: Indigenous Knowledge Systems and Sustainable Development–IDRC Program Development*. Ottawa: IDRC.

Latour, B. 1986. "Visualization and Cognition: Thinking with Eyes and Hands." In H. Kuklick and E. Long, eds., *Knowledge and Society: Studies in the Sociology of Culture Past and Present*. Vol. 6. Greenwich, Conn.: JAI Press.

Le Cointe, P. 1947. *Árvores e plantas uteis*. Rio de Janeiro: Companhia Editora Nacional.

Leach, M. 1994. *Rainforest Relations: Gender and Resource Use Among the Mende of Gola, Sierra Leone*. Washington, D.C.: Smithsonian Institution Press.

Leakey, R. E., and L. J. Slikkerveer. 1991. *Origins and Development of Agriculture in East Africa: The Ethnosystems Approach to the Study of Early Food Production in Kenya*. Studies in Technology and Social Change No. 19. Ames: CIKARD, Iowa State University.

Leisinger, K. 1995. "Sociopolitical Effects of New Biotechnologies in Developing Countries." IFPRI Discussion Paper No. 2. Washington, D.C.: IFPRI.

Leonard, H., ed. 1989. *Environment and the Poor: Development Strategies for a Common Agenda*. New Brunswick and Oxford: Transaction Books.

Leonard, J. 1987. *Natural Resources and Economic Development in Central America*. Washington, D.C.: IIED.

Lieth, H., and R. H. Whittaker, eds. 1975. *Primary Productivity of the Biosphere*. New York: Springer-Verlag.

Lightfoot, C. 1987. "Indigenous Research and On-Farm Trials." *Agricultural Administration and Extension Newsletter* 24:79–89.

Lightfoot, C., and R. Barker. 1988. "On-Farm Trials: A Survey of Methods." *Agricultural Administration and Extension Network* 30:15–23.

Long, N., and M. Villareal. 1994. "The Interweaving of Knowledge and Power in Development Interfaces." In I. Scoones and J. Thompson, eds., *Beyond Farmer First: Rural People's Knowledge, Agricultural Research and Extension Practice*. London: ITP.

Lutz, E., et al. 1993. "Interdisciplinary Fact-Finding on Current Deforestation in Costa Rica." Working Paper No. 61. Washington, D.C.: World Bank, Environment Department.

Lyotard, J. F. 1985. *The Post-Modern Connection*. Minneapolis: University of Minnesota.

Mahar, D. J. 1988. "Government Policies and Deforestation in Brazil's Amazon Region." Working Paper No. 7. Washington, D.C.: World Bank, Environment Department.

Marten, G., ed. 1986. *Traditional Agriculture in Southeast Asia: A Human Ecology Perspective.* Boulder, Colo.: Westview Press.

Mathias, E. 1994. *Indigenous Knowledge and Sustainable Development.* International Institute of Rural Reconstruction (IIRR) Working Paper No. 53. Silang, Cavite, Philippines: IIRR.

————. 1997. *Recording and Using Indigenous Knowledge: A Manual.* Silang, Cavite, Philippines: IIRR.

Mathias-Mundy, E. 1989. "Of Herbs and Healers." *ILEIA Newsletter* 4(3):20–22.

Mathias-Mundy, E., and C. M. McCorkle. 1989. *Ethnoveterinary Medicine: An Annotated Bibliography.* Bibliographies in Technology and Social Change No. 6. Ames: Technology and Social Change Program, Iowa State University.

Mathias-Mundy, E., O. Muchena, G. McKiernan, and P. Mundy. 1992. *Indigenous Technical Knowledge of Private Tree Management: A Bibliographic Report.* Bibliographies in Technology and Social Change No. 7. Ames: Technology and Social Change Program, Iowa State University.

Matlon, P., D. Cantrell, D. King, and M. Benoit-Cattin, eds. 1984. *Farmers' Participation in the Development of Technology: Coming Full Circle.* Ottawa: International Development Research Center.

Mattee, A. Z., and T. Lasalle. 1994. "Diverse and Limited: Farmers' Organizations in Tanzania." Agricultural Administration Research and Extension Network Paper No. 50. London: ODI.

Maurya, D., A. Bottrall, and J. Farrington. 1988. "Improved Livelihoods, Genetic Diversity and Farmer Participation: A Strategy for Rice Breeding in Rainfed Areas of India." *Experimental Agriculture* 24(3).

Mausolff, C., and S. Farber. 1995. "An Economic Analysis of Ecological Agricultural Technologies Among Peasant Farmers in Honduras." *Ecological Economics* 12:237–48.

McCall, M. M. 1995. *Indigenous Technical Knowledge in Farming Systems of East Africa: A Bibliography.* Bibliographies in Technology and Social Change No. 9. Revised edition. Ames: CIKARD, Iowa State University.

McClure, G. 1989. "Introduction." In D. M. Warren, L. J. Slikkerveer, and S. Oguntunji Titilola, eds., *Indigenous Knowledge Systems: Implications for Agriculture and International Development.* Studies in Technology and Social Change No. 11. Ames: Technology and Social Change Program, Iowa State University.

McCorkle, C. 1989. "Veterinary Anthropology." *Human Organization* 48(2): 156–62.

———. 1991. "Pooling Labor and Herds." *ILEIA Newsletter* 7(4):12–13.

———. 1994. *Farmer Innovation in Niger.* Studies in Technology and Social Change No. 21. Ames: Technology and Social Change Program, Iowa State University.

McCorkle, C. M., E. Mathias, and T. W. Schillhorn van Veen, eds. 1996. *Ethnoveterinary Research and Development.* London: ITP.

McCorkle, C. M., and G. McClure. 1995. "Farmer Know-How and Communication for Technology Transfer: CTTA in Niger." In D. M. Warren, L. J. Slikkerveer, and D. Brokensha, eds., *The Cultural Dimension of Development: Indigenous Knowledge Systems.* London: ITP.

McGrath, D. G., F. Castro, C. Futemma, B. D. Amaral, and J. Calabria. 1993. "Fisheries and the Evolution of Resource Management on the Lower Amazon Floodplain." *Human Ecology* 21(2):167–95.

McGregor, E., comp. 1994. *Indigenous and Local Community Knowledge in Animal Health and Production Systems–Gender Perspectives: A Working Guide to Issues, Networks and Initiatives.* Ottawa: World Women's Veterinary Association.

McMahon, M. 1994. *Getting Beyond the "National Institute Model" for Agricultural Research in Latin America: A Cross-Country Study of Brazil, Chile, Colombia and Mexico.* Latin America and the Caribbean Technical Department, Regional Studies Program, Washington, D.C.: World Bank.

Medellín-Morales, S. 1990. "Manejo agrosilvícola tradicional en una comunidad totonaca de la costa de Veracruz, México." In D. A. Posey and W. L. Overal, organizers, *Ethnobiology: Implications and Applications: Proceedings of the First International Congress of Ethnobiology.* Vol. 2. Belém, Brazil: Museu Paraense Emílio Goeldi.

Mellor, J., and G. Desai, eds. 1985. *Agricultural Change and Rural Poverty.* Baltimore, Md.: Johns Hopkins University Press.

Mendez. G. Jhonny. 1993. "Manejo del bosque natural en la región Huetar Norte de Costa Rica." *Revista Forestal Centroamericana* 6 (edición especial). Año 2:42–49.

Merrill-Sands D., S. D. Biggs, R. J. Bingen, P. T. Ewell, J. L. McAllister, and S. R. Poats. 1991. "Integrating On-Farm Research into National Agricultural Research Systems: Lessons for Research Policy, Organization and Management." In R. Tripp, ed., *Planned Change in Farming Systems: Progress in On-Farm Research.* Chichester, England: John Wiley and Sons.

Messerschmidt, D. 1991. "Community Forestry Management and the Opportunities of Local Traditions: A View from Nepal." In D. M. Warren, D. Brokensha, and L. J. Slikkerveer, eds., *Indigenous Knowledge Systems: The Cultural Dimension of Development*. London: Kegan Paul International.

Ministerio de Planificación Nacional y Política Económica. 1994. *Diagnóstico socioeconómico de la región Huetar Norte*. Costa Rica.

Moock, J. L., and R. E. Rhoades, eds. 1992. *Diversity, Farmer Knowledge, and Sustainability*. Ithaca, N.Y.: Cornell University Press.

Moris, J. 1991. *Extension Alternatives in Tropical Africa*. London: ODI.

Muchagata, M. G., V. de Reynal, and I. P. Verga Jr. 1994. "Building a Dialogue Between Researchers and Small Farmers: The Tocantins Agroecology Center (CAT) in Brazil." Agricultural Administration Research and Extension Network Paper 50d. London: ODI.

Mushita, T. A. 1993. "Strengthening the Informal Seed System in Communal Areas of Zimbabwe." In W. de Boef, K. Amanor, K. Wellard, and A. Bebbington, eds., *Cultivating Knowledge: Genetic Diversity, Farmer Experimentation and Crop Research*. London: ITP.

Natural Resources Defense Council. 1991. *Amazon Crude*. New York: Natural Resources Defense Council.

Niamir, M. 1990. *Community Forestry: Herders' Decision-Making in Natural Resources Management in Arid and Semi-Arid Africa*. Rome: FAO Community Forest Note No. 4.

———. 1995. "Indigenous Natural Resource Management Systems Among Pastoralists of Arid and Semi-Arid Africa." In D. M. Warren, L. J. Slikkerveer, and D. Brokensha, eds., *The Cultural Dimension of Development: Indigenous Knowledge Systems*. London: ITP.

Nielsen, N. O., ed. 1994. *Agroecosystem Health*. Proceedings of the International Workshop in Agroecosystem Health, University of Guelph, Guelph, Ontario. Published with the support of the International Development Research Center through the University of Guelph Agroecosystem Health Project.

Nimlos, T. J., and R. F. Savage. 1991. "Successful Soil Conservation in the Ecuadorian Highlands." *Journal of Soil and Water Conservation* (September–October):341–43.

Norem, R. H., R. Yoder, and Y. Martin. 1989. "Indigenous Agricultural Knowledge and Gender Issues in Third World Agricultural Development." In D. M. Warren, L. J. Slikkerveer, and S. Oguntunji Titilola, eds., *Indigenous Knowledge Systems: Implications for Agriculture and International Development*. Studies in Technology and Social Change Program No. 11. Ames: Iowa State University Research Foundation.

Bibliography

Norgaard, R. B. 1994. *Development Betrayed: The End of Progress and Co-Evolutionary Re-Envisioning of the Future.* London: Routledge.

Normann, H., I. Snyman, and M. Cohen, eds. 1996. *Indigenous Knowledge and Its Uses in Southern Africa.* Pretoria: Human Sciences Research Council.

NRC (National Research Council). 1989. *Alternative Agriculture.* Washington, D.C.: National Academy Press.

————. 1992a. *Conserving Biodiversity: A Research Agenda for Development Agencies.* Washington, D.C.: National Academy Press.

————. 1992b. *Neem: A Tree for Solving Global Problems.* Washington, D.C.: National Academy Press.

————. 1992c. *Vetiver Grass: A Thin Green Line Against Erosion.* Washington, D.C.: National Academy Press.

————. 1993a. *Managing Global Genetic Resources.* Washington, D.C.: National Academy of Sciences.

————. 1993b. *Sustainable Agriculture and the Environment in the Humid Tropics.* Washington, D.C.: National Academy of Sciences.

Obomsawim, R. 1993. *Culture Based Knowledge Systems in Development: Securing the Foundations for a Sustainable Future.* Consultancy Report submitted to the Canadian International Development Agency.

Office of Technology Assessment. 1984. *Technologies to Sustain Tropical Forest Resources.* Washington, D.C.: U.S. Congress.

Okali, C., V. Sumberg, and J. Farrington. 1994. *Battlefields and Trial Plots: Rhetoric and Reality of Farmer Participatory Research.* London: ITP.

Olson, J. S., J. A. Watts, and L. J. Allison. 1983. *Carbon in Live Vegetation of Major World Ecosystems.* Oak Ridge, Tenn.: Oak Ridge National Laboratory.

Orstrom, E. 1990. *Governing the Commons: The Evolution of Institutions for Collective Action.* Cambridge, England: Cambridge University Press.

Pachico D., J. A. Ashby, and L. R. Sanint. 1994. "Natural Resource and Agricultural Prospects for the Hillsides of Latin America." In *International Center for Tropical Agriculture (CIAT) Hillsides Program, Annual Report, 1993–1994.* Cali, Colombia: CIAT.

Pachico, D., and J. A. Ashby. 1983. "Stages in Technology Diffusion Among Small Farmers: Biological and Management Screening of a New Rice Variety in Nepal." *Agricultural Administration* 13:23–37.

Padoch, C. 1988. "Aguaje (*Mauritia flexuosa* L.F.) in the Economy of Iquitos, Peru." *Advances in Economic Botany* 6:214–24.

Padoch, C., and W. De Jong. 1991. "The House Gardens of Santa Rosa: Diversity and Variability in an Amazonian Agricultural System." *Economic Botany* 45(2):166–75.

Painter, M., and W. H. Durham. 1995. *The Social Causes of Environmental Destruction in Latin America.* Ann Arbor: University of Michigan Press.

Pandey, B., and S. Chaturvedi. 1993. "Indigenous Knowledge and Plant Genetic Resource Conservation in Ethiopia." *Biotechnology and Development Monitor* 16:19–20.

Pawluk, R. R., J. A. Sandor, and J. A. Tabor. 1992. "The Role of Indigenous Soil Knowledge in Agricultural Development." *Journal of Soil and Water Conservation* 47(4):298–302.

Pendelton, L. H. 1992. "Trouble in Paradise: Practical Obstacles to Nontimber Forestry in Latin America." In M. Plotkin and L. Famolare, eds., *Sustainable Harvest and Marketing of Rain Forest Products.* Washington, D.C.: Island Press.

Pérezgrovas, R., P. Pedraza, and M. Peralta. 1992. "Plants and Prayers: Animal Healthcare by Indian Shepherdesses." *ILEIA Newsletter* 8(3):22–23.

Picado, W. 1994. "Programa de incentivo al manejo de bosque secundario en Costa Rica (PIMBOS). Estructura de Costos. Informe de Consultoría." Unpublished document. Costa Rica.

Pichón, F. J., and J. E. Uquillas. 1995. "Rural Poverty Alleviation and Improved Natural Resource Management Through Participatory Technology Development in Latin America's Risk-Prone Areas." Paper presented at the Workshop on Traditional and Modern Approaches to Natural Resource Management in Latin America and the Caribbean, 25–26 April 1995, World Bank. Washington, D.C.

———. 1996. *Sustainable Agriculture and Poverty Reduction in Latin America's Risk-Prone Areas: Opportunities and Challenges.* Latin America and the Caribbean Technical Department, Regional Studies Program Report No. 40. Washington, D.C.: World Bank.

———. 1997. "Agricultural Intensification and Poverty Reduction in Latin America's Risk-Prone Areas." *Journal of Developing Areas* 31(4):479–514.

———. 1998. "Sustainable Agriculture Through Farmer Participation: Agricultural Research and Technology Development in Latin America's Risk-Prone Areas." In J. Blauert and S. Zadek, eds., *Mediating Sustainability: Growing Policy from the Grassroots.* West Hartford, Conn.: Kumarian Press.

Pimentel, D., and M. Pimentel. 1979. *Food, Energy and Society.* London: Arnold.

Pingali, P., and H. Binswanger. 1987. "Population Density and Agricultural Intensification: A Study of the Evolution of Technologies in Tropical Agriculture." In G. D. Johnson and R. D. Lee, eds., *Population Growth and Economic Development: Issues and Evidence.* Madison: University of Wisconsin Press.

Pinstrup-Andersen, P., and R. Pandya-Lorch. 1995. "Alleviating Poverty, Intensifying Agriculture, and Effectively Managing Natural Resources." IFPRI Discussion Paper No. 1. Washington, D.C.: IFPRI.

Pinstrup-Andersen, P., R. Pandya-Lorch, and P. Hazell. 1994. "Hunger, Food Security, and the Role of Agriculture in Developing Countries." Washington, D.C.: IFPRI.

Poffenberger, M., ed. 1990. *Keepers of the Forest: Land Management Alternatives in Southeast Asia.* West Hartford, Conn.: Kumarian Press.

Portillo, H., H. Pitre, D. Meckenstock, and K. Andrews. 1991. "Langosta: A Lepidopterous Pest Complex on Sorghum and Maize in Honduras." *Florida Entomologist* 74:288–96.

Posey, D. A. 1982. "The Keepers of the Forest." *Garden* 6(1):18–24.

———. 1983. "Indigenous Ecological Knowledge and Development of the Amazon." In E. F. Moran, ed., *Dilemma of Amazonian Development.* Boulder, Colo.: Westview Press.

———. 1985. "Indigenous Management of Tropical Forest Ecosystems: The Case of the Kayapó Indians of the Brazilian Amazon." *Agroforestry Systems* 3:139–58.

———. 1991. "Kayapó Indians: Experts in Synergy." *ILEIA Newsletter* 7(4):3–5.

Posey, D. A., and W. Balee, eds. 1989. *Resource Management in Amazonia: Indigenous and Folk Strategies.* Advances in Economic Botany No. 7. Bronx, N.Y.: New York Botanical Garden.

Posey, D. A., J. Frechione, J. Eddins, L. F. Da Silva, D. Myers, D. Case, and P. Macbeth. 1984. "Ethnoecology as Applied Anthropology in Amazonian Development." *Human Organization* 43(2):95–107.

Poveda, J. 1991. "La Amazonía: Conflicto de esperanzas y realidades." Unpublished manuscript.

Prain, G., and C. P. Bagalanon, eds. 1994. *Local Knowledge, Global Science and Plant Genetic Resources: Toward a Partnership.* Los Banos, Calif.: UPWARD.

Prain, G.D., S. Fujisaka, and D.M. Warren, eds. 1999. *Biological and Cultural Diversity: The Role of Indigenous Agricultural Experimentation in Development.* London: ITP.

Prain, G., F. Uribe, and U. Scheidegger. 1991. "Small Farmers in Agricultural Research: Farmer Participation in Potato Germplasm Evaluation." In B. Haverkort, J. van der Kamp, and A. Waters-Bayer, eds., *Joining Farmers' Experiments: Experiences in Participatory Technology Development.* London: ITP.

Prescott-Allen, C., and R. Prescott-Allen. 1986. *The First Resource: Wild Species in the North American Economy.* New Haven and London: Yale University Press.

Pretty, J. N. 1991. "Farmers' Extension Practice and Technology Adaptation: Agricultural Revolution in Seventeenth–Nineteenth Century Britain." *Agriculture and Human Values* 8(1–2):132–48.

————. 1995. *Regenerating Agriculture: Policies and Practice for Sustainability and Self-Reliance.* London: Earthscan.

Psacharopoulos, G., and H. Patrinos, eds. 1994. *Indigenous People and Poverty in Latin America: An Empirical Analysis.* Washington, D.C.: World Bank, World Bank Regional and Sectoral Studies.

Quintana, J. 1992. "American Indian Systems for Natural Resource Management." *Akwe:kon Journal* 9(2):92–97.

Quirós, C. A., T. Gracia, and J. A. Ashby. 1991. *Farmer Evaluations of Technology: Methodology for Open-Ended Evaluation.* Instructional Unit No. 1. CIAT, Palmira. Also in Spanish.

Quiros, D., and C. Reiche. 1995. *Manejo sustentable de un bosque natural tropical en Costa Rica: Análisis financiero.* Serie Técnica No. 11. Colección Silvicultura y Manejo de Bosques Naturales. Costa Rica.

Quiroz, C. 1993. "Methodology for the Study of Farmers' Agricultural Local 'Indigenous' Knowledge Systems in Rural Development Programs: In-Service Training Program for Research/Extension Practitioners—An Experience from Venezuela." In *Indigenous Knowledge and Sustainable Development.* Twenty-five selected papers presented at the International Symposium on Indigenous Knowledge and Sustainable Development. Silang, Cavite, Philippines: IIRR.

————. 1994. "Biodiversity, Indigenous Knowledge, Gender and Intellectual Property Rights." *Indigenous Knowledge and Development Monitor* (special issue) 2(3):12–15.

————. 1995. "Farmer Experimentation in a Venezuelan Andean Region." In D. M. Warren, S. Fujisaka, and G. Prain, eds., *Indigenous Experimentation and Cultural Diversity.* London: ITP.

Rajasekaran, B. 1994. *A Framework for Incorporating Indigenous Knowledge Systems into Agricultural Research, Extension and NGOs for Sustainable Agricultural Development.* Studies in Technology and Social Change No. 22. Ames: Technology and Social Change Program, Iowa State University.

Rajasekaran, B., and D. M. Warren. 1991. "Indigenous Rice Taxonomies and Farmers' Rice Production Decision-Making Systems in South India." In D.M. Warren, D. Brokensha, and L.J. Slikkerveer, eds., *Indigenous Knowledge Systems: The Cultural Dimension of Development,* London: Kegan Paul International.

Rajasekaran, B., D. M. Warren, and S. C. Babu. 1991. "Indigenous Natural-Resource Management Systems for Sustainable Agricultural Development—A Global Perspective." *Journal of International Development* 3(4):387–401.

Rau, B. 1991. *From Feast to Famine: Official Cures and Grassroots Remedies to Africa's Food Crisis.* London: Zed Books.

Ravnborg, H. 1994. "Methodology for Identifying Different Levels of Well-Being, and Their Distribution." International Center for Tropical Agriculture (CIAT) Hillsides Program Internal Discussion Paper. Palmira: CIAT.

Rechkemmer, P. 1996. *Women Farmers of Ara, Nigeria.* Masters thesis, Department of Anthropology, Iowa State University, Ames, Iowa.

Redford, K. H., and C. Padoch. 1992. *Conservation in Neotropical Forests: Working from Traditional Resource Use.* New York: Columbia University Press.

Reichel, E. 1992. "Shamanistic Modes for Environmental Accounting in the Colombian Amazon: Lessons from Indigenous Ethno-Ecology for Sustainable Development." Paper presented at the International Symposium on Indigenous Knowledge and Sustainable Development. Silang, Cavite, Philippines: IIRR.

Reichhardt, K. L., E. Mellink, G. P. Nabhan, and A. Rea. 1994. "Habitat Heterogeneity and Biodiversity Associated with Indigenous Agriculture in the Sonoran Desert." *Etnoecologica* 2(3):21–34.

Reij, C. 1993. "Improving Indigenous Soil and Water Conservation Techniques: Does It Work?" *Indigenous Knowledge and Development Monitor* 1(1):11–13.

Reijntjes, C. 1991. "Raised Fields for Lowland Farming." *ILEIA Newsletter* 7(4):6–8.

Reijntjes, C., B. Haverkort, and A. Waters-Bayer. 1992. *Farming for the Future: An Introduction to Low-External-Input and Sustainable Agriculture.* London: Macmillan Press.

Repetto, R. 1985. "Paying the Price: Pesticide Subsidies in Developing Countries." World Resources Institute Research Report No. 2. Washington, D.C.: WRI.

REPPIKA (Regional Program for the Promotion of Indigenous Knowledge in Asia). 1994. *Indigenous Knowledge and Sustainable Development in the Philippines.* Silang, Cavite, Philippines: IIRR.

Rhoades, R. 1984. *Breaking New Ground: Agricultural Anthropology.* Lima: CIP.

———. 1987. "Farmers and Experimentation." Agricultural Administration and Extension Network Paper No. 21. London: ODI.

———. 1988. "Changing Perceptions of Farmers and the Expanding Challenges of International Agricultural Research." Paper presented at the

Conference of Farmers and Food Systems, 26–30 September 1988, Lima, Peru.

Rhoades, R., and A. Bebbington. 1995. "Farmers Who Experiment: An Untapped Resource for Agricultural Research and Development." In D. M. Warren, S. Fujisaka, and G. Prain, eds., *Indigenous Experimentation and Cultural Diversity.* London: ITP.

Rhoades, R., and R. Booth. 1982. "Farmer-Back-to-Farmer: A Model for Generating Acceptable Agricultural Technology." *Agricultural Administration* 11:127–37.

Richards, P. 1985. *Indigenous Agricultural Revolution: Ecology and Food Production in West Africa.* London: Hutchinson.

———. 1989. "Agriculture as a Performance." In R. Chambers, A. Pacey, and L. A. Thrupp, eds., *Farmer First: Farmer Innovations and Agricultural Research.* London: ITP.

Rist, S. 1991a. "Participation, Indigenous Knowledge and Trees." *Forests, Trees and People Newsletter* 13:30–36.

———. 1991b. "Revitalizing Indigenous Knowledge." *ILEIA Newsletter* 7(3): 22–23.

———. 1994. "The Aynokas: Sustaining Agri-Culture." *ILEIA Newsletter* 10(2):6–8.

Rivera-Cucicanqui, S. 1990. "Liberal Democracy and Ayllu Democracy in Bolivia: The Case of Northern Potosí." In J. Fox, ed., *The Challenge of Rural Democratization: Perspectives from Latin America and the Philippines.* London: Frank Cass.

Rocheleau, D. E. 1991. "Gender, Ecology, and the Science of Survival: Stories and Lessons from Kenya." *Agriculture and Human Values* 8(1–2):156–65.

Rogers, E. M. 1969. *Modernization Among Peasants: The Impact of Communication.* New York: Holt, Rinehart and Winston.

———. 1983. *Diffusion of Innovation.* 3d edition. New York: Free Press.

Rola, A., and P. Pingali. 1993. *Pesticides, Rice Productivity, and Farmers' Health: An Economic Assessment.* Manila and Washington, D.C.: IIRR and WRI.

Roling, H. 1988. *Extension Science: Information Systems in Agricultural Development.* Cambridge, England: Cambridge University Press.

Rolling, N., and P. Engel. 1989. "IKS and Knowledge Management: Utilizing Indigenous Knowledge in Institutional Knowledge Systems." In D. M. Warren, L. J. Slikkerveer, and S. Oguntunji Titilola, eds., *Indigenous Knowledge Systems: Implications for Agriculture and International Development.* Studies in Technology and Social Change Program No. 11. Ames: Iowa State University Research Foundation.

Romanoff, S. 1990. "On Reducing the Cost of Promoting Local Farmer Organizations in Agricultural Development Projects." Paper presented at the Annual Farming Systems Research and Extension Symposium, East Lansing, Michigan.

———. 1993. "Farmers' Organization, Research and Diffusion of Technology." In K. Dvorzak, ed., *Social Science Research for Agricultural Technology Development.* Wallingford, U.K.: CABI.

Rosegrant, M., and M. Svendsen. 1993. "Asian Food Production in the 1990s: Irrigation Investment and Management Policy." *Food Policy* 18 (February):13–32.

Ruddell, E. 1994. *Documentation of Farmer Experiments: A Key Strategy for Achieving Food Security on a Sustainable Basis.* Santiago, Chile: Vecinos Mundiales.

———. 1995. "Growing Food for Thought: A New Model of Site-Specific Research from Bolivia." *Grassroots Development* 19(1):18–26.

Ruttan, V., and Y. Hayami. 1990. "Induced Innovation Model of Agricultural Development." In C. Eicher and J. Staatz, eds., *Agricultural Development in the Third World.* 2d edition. Baltimore, Md.: Johns Hopkins University Press.

Salas, M. A. 1994. "The Technicians Only Believe in Science and Cannot Read the Sky: The Cultural Dimension of the Knowledge Conflict in the Andes." In I. Scoones and J. Thompson, eds., *Beyond Farmer First.* London: ITP.

Salas, M., and T. Tillman. 1990. "Peasants Are Proud to Share Their Knowledge." *Forests, Trees and People Newsletter* 9–10:44.

SANREM (Sustainable Agricultural and Natural Resource Management). 1993. *SANREM Ecolinks.* Quarterly newsletter of the Sustainable Agriculture and Natural Resource Management Collaborative Research Support Program, University of Georgia, Athens, Georgia.

Scherr, S., and P. Hazell. 1993. "Sustainable Agricultural Development Strategies in Fragile Lands." Paper prepared for the American Agricultural Economics Association's 1993 International Pre-Conference on Post–Green Revolution Agricultural Development Strategies in the Third World: What Next? 30–31 July 1993, Orlando, Florida.

Schultes, R. E. 1980. "The Amazonia as a Source of New Economic Plants." *Economic Botany* 33(3):259–66.

———. 1990. "Gifts of the Amazon Flora to the World: Many Valuable Plants Were Originally Domesticated in the Now Threatened Rain Forest of the Amazon Basin." *Arnoldia* 50(2):21–34.

Schwartz, L. A. 1994. "The Role of the Private Sector in Agricultural Extension Economic Analysis and Case Studies." Agricultural Administration Research and Extension Network Paper No. 48. London: ODI.

Scoones, I., and J. Thompson, eds. 1994. *Beyond Farmer First: Rural People's Knowledge, Agricultural Research and Extension Practice.* London: ITP and IIED.

Shah, P. 1994. "Village-Managed Extension Systems in India: Implications for Policy and Practice." In I. Scoones and J. Thompson, eds., *Beyond Farmer First: Rural People's Knowledge, Agricultural Research and Extension Practice.* London: ITP and IIED.

Shaw, R., G. Gallopín, P. Weaver, and S. Öberg. 1992. *Sustainable Development: A Systems Approach.* Status Report SR-92-6.

Sherwood, S., and J. Bentley. 1995. "Rural Farmers Explore Causes of Plant Disease." *ILEIA Newsletter* 11(1):20–22.

Shiva, V. 1993. *Monocultures of the Mind: Perspectives on Biodiversity and Biotechnology.* London and New Jersey: Zed Books, Third World Network.

Shiva, V., and I. Dankelman. 1992. "Women and Biological Diversity: Lessons from the Indian Himalaya." In D. Cooper, R. Vellve, and H. Hobbelink, eds., *Growing Diversity: Genetic Resources and Local Food Security.* London: ITP.

Simpson, B. M. 1994. "Gender and the Social Differentiation of Local Knowledge." *Indigenous Knowledge and Development Monitor* (special issue) 2(3): 21–23.

Slaybaugh-Mitchell, T. L. 1995. *Indigenous Livestock Production and Husbandry: An Annotated Bibliography.* Bibliographies in Technology and Social Change No. 8. Ames: CIKARD, Iowa State University.

Smith, N. J. H. 1981. *Man, Fishes, and the Amazon.* New York: Columbia University Press.

———. 1995. "Biodiversity and Agroforestry Along the Amazon Floodplain." Paper presented at the Workshop on Traditional and Modern Approaches to Natural Resource Management in Latin America and the Caribbean, 25–26 April 1995, World Bank, Washington, D.C.

———. 1996. "Home Gardens as a Springboard for Agroforestry Development in Amazonia." *International Tree Crops Journal* 9:11–30.

———. *The Amazon River Forest: A Natural History of Plants, Animals, and People.* New York: Oxford University Press.

Smith, N. J. H., I. C. Falesi, P. Alvim, and E. A. Serrão. 1995. "Agroforestry Development and Potential in the Brazilian Amazon." *Land Degradation and Rehabilitation* 6:251–263.

Smith, N. J. H., and R. E. Schultes. 1990. "Deforestation and Shrinking Crop Genepools in Amazonia." *Environmental Conservation* 17(3):227–34.

Smith, N. J. H., J. T. Williams, D. L. Plucknett, and J. P. Talbot. 1992. *Tropical Forests and Their Crops.* Ithaca, N.Y.: Cornell University Press.

Solís, M., and C. Reiche. 1995. *Experiencias técnicas, económicas y de participación en el manejo sostenible de los bosques de COOPESANJUAN R.L. Región Huetar Norte, Costa Rica.* Estudio de caso presentado en el II Congreso Forestal Centroamericano, Honduras.

Soria, C. 1995. Personal communication with J. Ashby and T. Gracia about Bolivian CIALs during 1993–1994.

Southgate, D. 1991. "Tropical Deforestation and Agricultural Development in Latin America." World Bank Environment Department Policy Research Division Working Paper No. 20. Washington, D.C.: World Bank.

———. 1992. "Promoting Resource Degradation in Latin America: Tropical Deforestation, Soil Erosion and Coastal Ecosystem Disturbance in Ecuador." *Economic Development and Cultural Change* 40(4):787–807.

Southgate, D., and C. Runge. 1990. "The Institutional Origins of Deforestation in Latin America." Department of Agricultural and Applied Economics, University of Minnesota. Staff Paper P90-S.

Sperling, L., U. Scheidegger, R. Buruchara, P. Nyabyenda, and S. Munyanesa. 1994. "Intensifying Production Among Smallholder Farmers: The Impact of Improved Climbing Beans in Rwanda." CIAT Network on Bean Research in Africa Occasional Publications No. 12. Butare, Rwanda: CIAT/ RESAPAC.

Spittler, M. P. 1995. Inventarios en bosques secundarios en la Zona Norte de Costa Rica. Personal communication.

Spruce, R. 1908. *Notes of a Botanist on the Amazon and Andes.* London: Macmillan.

Stearman, A. M. 1994. "Revisiting the Myth of the Ecologically Noble Savage in Amazonia: Implications for Indigenous Land Rights." *Culture and Agriculture* 49:2–6.

Sternberg, H. O. "Waters and Wetlands of Brazilian Amazonia: An Uncertain Future." In T. Nishizawa and J. I. Uitto, eds., *The Fragile Tropics of Latin America: Sustainable Management of Changing Environments.* Tokyo: United Nations University Press.

Stolzenbach, A. 1994. "Learning by Improvisation: Farmers' Experimentation in Mali." In I. Scoones and J. Thompson, eds., *Beyond Farmer First: Rural People's Knowledge, Agricultural Research and Extension Practice.* London: ITP and IIED.

Sumberg, J., and C. Okali. 1997. *Farmers' Experiments: Creating Local Knowledge.* Boulder, Colo.: Lynne Reinner.

TAC/CGIAR (Technical Advisory Committee/Consultative Group on International Agricultural Research). 1988. *Sustainable Agricultural Production: Implications for International Agricultural Research.* Rome: FAO.

———. 1993. *The Ecoregional Approach to Research in the CGIAR: Report of the TAC/Center Directors Working Group.* Rome: FAO, TAC Secretariat.

Tapia, M. E., and M. Banegas. 1990. "Human Adaptation to a High-Risk Environment: Camellones or Waru Waru—Traditional Agricultural Technology of the Peruvian Andes." *Journal of Farming Systems Research-Extension* 1(1):93–98.

Tendler, J., with K. Healey and C. C. N. O'Laughlin. 1988. "What to Think About Cooperatives: A Guide from Bolivia." In S. Annis and P. Hakim, eds., *Direct to the Poor: Grassroots Development in Latin America.* London: Lynne Reinner. Also published as a larger report by the Inter-American Foundaton, Rosslyn, Virginia.

Thrupp, L. 1989. "Legitimizing Local Knowledge: From Displacement to Empowerment for Third World People." *Agriculture and Human Values* 6(3): 13–24.

Thurston, H. D. 1992. *Sustainable Practices for Plant Disease Management in Traditional Farming Systems.* Boulder, Colo.: Westview Press.

Thurston, H. D., and J. M. Parker. 1995. "Raised Beds and Plant Disease Management." In D. M. Warren, S. Fujisaka, and G. Prain, eds., *Indigenous Experimentation and Cultural Diversity.* London: ITP.

Tick, A., ed. 1993. *Indigenous Knowledge and Development Monitor.* Published three times a year by the Center for International Research and Advisory Networks (CIRAN), The Hague, the Netherlands.

Titilola, S. O. 1990. *The Economics of Incorporating Indigenous Knowledge Systems into Agricultural Development: A Model and Analytical Framework.* Studies in Technology and Social Change No. 17. Ames: Technology and Social Change Program, Iowa State University.

Titilola, S., L. Egunjobi, A. Amusan, and B. Wahab. 1995. *Introduction of Indigenous Knowledge into the Education Curriculum of Primary, Secondary and Tertiary Institutions in Nigeria: A Policy Guideline.* Ames: CIKARD, Iowa State University.

Toledo, V. M., B. Ortiz, and S. Medellín-Morales. 1994. "Biodiversity Islands in a Sea of Pasturelands: Indigenous Resource Management in the Humid Tropics of Mexico." *Etnoecologica* 2(3):37–49.

Torné de Valcárel, F. 1991. "The Rural University Belongs to the People." *ILEIA Newsletter* 7(3):14–15.

Tripp, R., ed. 1991. *Planned Change in Farming Systems: Progress in On-Farm Research.* Chichester, England: John Wiley and Sons.

Trujillo, J. 1993. "El Ceibo." In A. Bebbington, G. Thiele, P. Davies, M. Prager, and H. Riveros, eds., *Non-Governmental Organizations and the State in Latin America.* London: Routledge.

Tulchin, J., ed. 1991. *Economic Development and Environmental Protection in Latin America.* Boulder, Colo.: Lynne Rienner.

UN (United Nations). 1984. *Population, Resources, Environment and Development.* Proceedings of expert group meeting, 25–29 April 1984, United Nations, Geneva.

———. 1985. *World Demographics Estimates, 1950–2050.* New York: UN.

———. 1989. *World Population Trends and Policies: 1987 Monitoring Report.* New York: UN.

UNEP (United Nations Environment Program). 1994. "Intergovernmental Committee on the Convention of Biological Diversity." 2d sess., UNEP-CBD-IC-2-14. Nairobi: UNEP.

UNESCO (United Nations Education, Scientific and Cultural Organization). 1994a. "Traditional Knowledge in Tropical Environments." *Nature and Resources* (special issue) 30(1).

———. 1994b. "Traditional Knowledge into the Twenty-First Century." *Nature and Resources* (special issue) 30(2).

UNFPA (United Nations Population Fund). 1990. *The State of World Population 1990.* New York: UN.

Uphoff, N. T. 1985. "Fitting Projects to People." In M. M. Cernea, ed., *Putting People First: Sociological Variables in Rural Development.* New York: Oxford University Press.

———. 1992. *Learning from Gal Oya: Possibilities for Participatory Development and Post-Newtonian Social Science.* Ithaca, N.Y.: Cornell University Press.

Uquillas, J. 1984. "Colonization and Spontaneous Settlement in the Ecuadorian Amazon." In M. Schmink and C. Wood, eds., *Frontier Expansion in Amazonia.* Gainesville: University of Florida Press.

———. 1993. "Research and Extension Practice and Rural People's Agroforestry Knowledge in Ecuadorian Amazonia." *International Institute for Environment and Development Research Series* 1(4).

Uquillas, J., and F. Pichón. 1995. *Rural Poverty Alleviation and Improved Natural Resource Management Through Participatory Technology Development in Latin America's Risk-Prone Areas.* Washington, D.C.: World Bank Technical Environmental Unit, Latin America and Caribbean Region.

van Veldhuizen, L., A. Waters-Bayer, R. Ramirez, D. Johnson, and J. Thompson, eds. 1997. *Farmers' Experimentation in Practice: Lessons from the Field.* London: ITP.

Van Niekerk, N. 1994. *Desarrollo rural en los Andes: Un estudio sobre los programas de organizaciones no gubernamentales.* Leiden, the Netherlands: Leiden Development Studies No. 13.

Van Crowder, L. 1991. "Extension for Profit." *Human Organization* 50(1):39–42.

Vanek, E. 1989. "Enhancing Resource Management in Developing Nations Through Improved Attitudes Towards Indigenous Knowledge Systems: The Case of the World Bank." In D. M. Warren, L. J. Slikkerveer, and S. Oguntunji Titilola, eds., *Indigenous Knowledge Systems: Implications for Agriculture and International Development.* Studies in Technology and Social Change Program No. 11. Ames: Iowa State University Research Foundation.

Vargas, R. 1995. DGF, personal communication.

Verhelst, T. 1990. *No Life Without Roots: Culture and Development.* London: Zed Books.

Villavicencio, R., R. Aquino, and J. Aquino. 1993. "Independencia Women Seek Independence: Making Andean Noodles." *ILEIA Newsletter* 9(3):24–25.

von Liebenstein, G. W. 1994. "The Monitor: The Future of an Instrument for Networking." *Indigenous Knowledge and Development Monitor* 2(2):2–3.

von Liebenstein, G. W., and A. W. Tick. 1994. "Indigenous Knowledge: Towards an Effective Strategy." Paper prepared for the Sri Lanka National Symposium on Indigenous Knowledge and Sustainable Development. The Hague: CIRAN.

Wachter, D., and J. English. 1992. *The World Bank's Experience with Rural Land Titling.* Environmental Policy and Research Divisional Working Paper. Washington, D.C.: World Bank.

Warren, D. M. 1987. "Utilizing Indigenous Healers in National Health Delivery Systems: The Ghanaian Experience." In J. van Willigen, B. Rylko-Bauer, and A. McElroy, eds., *Making Our Research Useful: Case Studies in the Utilization of Anthropological Knowledge.* Boulder, Colo.: Westview Press.

———. 1989. "Linking Scientific and Indigenous Agricultural Systems." In J. L. Compton, ed., *The Transformation of International Agricultural Research and Development.* Boulder, Colo.: Lynne Rienner.

———. 1991a. *Using Indigenous Knowledge in Agricultural Development.* World Bank Discussion Paper No. 127. Washington, D.C.: World Bank.

———. Guest ed. 1991b. *Indigenous Agricultural Knowledge Systems and Development.* Special issue of *Agriculture and Human Values* 8(1–2).

———. 1992. "Indigenous Knowledge, Biodiversity Conservation and Development." Keynote address presented at the International Conference on Conservation and Biodiversity in Africa—Local Initiatives and Institutional Roles. National Museum of Kenya, Nairobi.

———. 1994. "Indigenous Agricultural Knowledge, Technology, and Social Change." In G. McIsaac and W. Edwards, eds., *Sustainable Agriculture in the American Midwest: Lessons from the Past, Prospects for the Future.* Urbana: University of Illinois Press.

———. 1995a. "Indigenous Knowledge Systems for Sustainable Agriculture in Africa." In V. Udoh James, ed., *Sustainable Development in Third World Countries: Applied and Theoretical Perspectives.* Westport, Conn.: Greenwood.

———. 1995b. "Indigenous Knowledge and Sustainable Agricultural and Rural Development: Policy Issues and Strategies for the INDISCO Program of the ILO." In *Proceedings of the INDISCO Technical Review Meeting, Chiang Mai, Thailand.* Geneva: Cooperative Branch, International Labor Office (ILO).

———. 1995c. "Indigenous Knowledge, Biodiversity Conservation and Development: A Keynote Address." In L. Bennun, R. A. Aman, and S. A. Crafter, eds., *Conservation of Biodiversity in Africa: Local Initiatives and Institutional Roles.* Nairobi: Center for Biodiversity, National Museum of Kenya.

———. 1995d. "Using Indigenous Knowledge in Agricultural Development." Paper presented at the Workshop on Traditional and Modern Approaches to Natural Resource Management in Latin America and the Caribbean, 25–26 April 1995, World Bank, Washington, D.C.

Warren, D. M., and K. Cashman. 1989. *Indigenous Knowledge for Sustainable Agricultural and Rural Development.* Gatekeeper Series Paper No. 10. London: IIED.

Warren, D. M., L. Egunjobi, and B. Wahab, eds. 1996. *Indigenous Knowledge in Education.* Ibadan: Indigenous Knowledge Study Group, University of Ibadan, Nigeria.

Warren, D. M., S. Fujisaka, and G. Prain, eds. 1995. *Indigenous Experimentation and Cultural Diversity.* London: ITP.

Warren, D. M., and J. Pinkston. 1997. "Indigenous African Resource Management of a Tropical Rainforest Ecosystem: A Case Study of the Yoruba of Ara, Nigeria." In F. Berkes and C. Folke, eds., *Linking Social and Ecological Systems.* Cambridge, England: Cambridge University Press.

Warren, D. M., and B. Rajasekaran. 1993. "Indigenous Knowledge: Putting Local Knowledge to Good Use." *International Agricultural Development* 13(4):8–10.

———. 1995. "Using Indigenous Knowledge for Sustainable Dry-Land Management: A Global Perspective." In D. Stiles, ed., *Social Aspects of Sustainable Dryland Management.* Chichester, England: John Wiley and Sons, on behalf of UNEP.

Warren, D. M., L. J. Slikkerveer, and D. Brokensha, eds. 1995. *The Cultural Dimension of Development: Indigenous Knowledge Systems.* London: ITP.

Warren, D.M., S.O. Titilola, and L.J. Slikkerveer, eds. 1989. *Indigenous Knowledge Systems: Implications for Agriculture and International Development.* Studies in Technology and Social Change No. 11. Ames: Technology and Social Change Program, Iowa State University.

Warren, D. M., and M. S. Warren. 1999. "Local-Level Experimentation with Social Organization and Management of Self-Reliant Agricultural Development: The Case of Ara, Nigeria." In G. Prain, S. Fujisaka, and D. M. Warren, eds., *Biological and Cultural Diversity: The Role of Indigenous Agricultural Experimentation in Development.* London: ITP.

WCMC (World Conservation Monitoring Center). 1992. *Global Biodiversity: Status of the Earth's Living Resources.* London: Chapman and Hall.

Weber, P. 1992. "A Place for Pesticides?" *Worldwatch* 15(3):22–23.

WHO (World Health Organization). 1980. *Disease Prevention and Control in Water Development Schemes.* Geneva: WHO.

———. 1983. *Environmental Health Impact Assessment of Irrigated Agricultural Development Projects.* Geneva: WHO.

Wiens, T. 1994. *Rural Poverty, Sustainable Natural Resource Management, and Overall Rural Development in the LAC Region: The World Bank's Strategy.* Washington, D.C.: World Bank, Latin America and the Caribbean Technical Department, Background Note.

Wiggins, S., and E. Cromwell. 1995. "NGOs and Seed Provision to Smallholders in Developing Countries." *World Development* 23(3):413–22.

Wilcox, B. A. 1994. *Ecosystem Goods and Services: An Operational Framework.* The 2050 Project. Washington, D.C.: WRI.

———. 1995. "Rural Development and Indigenous Resources: Toward a Geographic-Based Assessment Framework." Paper presented at the Workshop on Traditional and Modern Approaches to Natural Resource Management in Latin America and the Caribbean, 25–26 April 1995, World Bank, Washington, D.C.

Wilcox, B. A., and K. N. Duin. 1994. *Biodiversity and Utility: Application to Ecoregions of Latin America and the Caribbean.* Report presented at the Biodiversity Support Program Workshop on Geographic Biodiversity Investment Priorities, 27 September–1 October 1994, Miami, Florida.

———. 1995. "Indigenous Cultural and Biological Diversity: Overlapping

Values of Latin America Ecoregions." *Cultural Survival Quarterly* 18(4): 49–53.

Wilcox, B. A., K. S. Smallwood, and J. R. Kahn. 1999. "Toward a P-S-R Indicator Approach for Forest Ecosystem Health and Natural Capital." Paper presented at the International Congress on Ecosystem Health, Sacramento, California.

Wilken, G. 1987. *Good Farmers: Traditional Agricultural Resource Management in Mexico and Central America.* Berkeley and Los Angeles: University of California Press.

Wilson, E. O. 1992. *The Diversity of Life.* New York: W. W. Norton.

Winslow, D. 1996. *An Annotated Bibliography of Naturalized Knowledge Systems in Canada.* Bibliographies in Technology and Social Change No. 10. Ames: CIKARD, Iowa State University.

Wolf, E. 1986. *Beyond the Green Revolution: New Approaches for Third World Agriculture.* Washington, D.C.: Worldwatch Institute.

Worede, M., and H. Mekbib. 1993. "Linking Genetic Resource Conservation to Farmers in Ethiopia." In W. de Boef, K. Amanor, K. Wellard, and A. Bebbington, eds., *Cultivating Knowledge: Genetic Diversity, Farmer Experimentation and Crop Research.* London: ITP.

World Bank. 1989. *Latin America: New Directions for Agricultural Credit Projects and Rural Financial Policies.* Washington, D.C.: World Bank, Latin America and the Caribbean Technical Department, Trade, Finance and Industry Division.

———. 1990. *World Development Report.* Washington, D.C.: World Bank.

———. 1991. *World Development Report.* Washington, D.C.: World Bank.

———. 1992. *World Development Report.* Washington, D.C.: World Bank.

———. 1993a. *Towards a Bank Strategy for Agriculture in the Latin America and Caribbean Region.* Washington, D.C.: World Bank, Latin America and the Caribbean Technical Department.

———. 1993b. *Agricultural Sector Review.* Washington, D.C.: World Bank, Agriculture and Natural Resources Department.

———. 1994. *World Development Report.* Washington, D.C.: World Bank.

World Bank/AGRTN (Agricultural Technology Note, Central Agricultural Department). 1994a. "Integrated Pest Management: An Environmentally Sustainable Approach to Crop Protection." Agriculture Technology Note No. 2. Washington, D.C.: World Bank.

———. 1994b. "Integrated Soil Management for the Tropics: A Profitable and Environmentally Sound Approach to Land Husbandry." Agriculture Technology Note No. 7. Washington, D.C.: World Bank.

————. 1995a. "Biotechnology: New Tools for Agriculture and the Environment." Agriculture Technology Note No. 8. Washington, D.C.: World Bank.

————. 1995b. "Competitive Research Grant Systems: An Approach to Funding Agricultural Research and Technology Development." Agriculture Technology Note No. 9. Washington, D.C.: World Bank.

World Commission on Environment and Development. 1987. *Our Common Future.* Oxford: Oxford University Press.

WRI (World Resources Institute). 1989. *World Resources 1988–89: An Assessment of the Resource Base That Supports the Global Economy.* New York: Basic Books.

————. 1990. *World Resources 1990–91: A Guide to the Global Environment.* Oxford: Oxford University Press.

————. 1992. *World Resources 1992–93: A Guide to the Global Environment.* Oxford: Oxford University Press.

WRI/CIDE (World Resources Institute/Center for International Development and Environment). 1990. *Directory of Country Environmental Studies.* Washington, D.C.: WRI.

WRI/CIDE and USAID/LAC (U.S. Agency for International Development, Bureau of Latin America and the Caribbean). 1991. *Environmental Strategy for Latin America and the Caribbean.* Washington, D.C.

Wright, N. G., ed. 1995. *Traditional Wisdom and Modern Know-How: Report of the Consultation on Fusion of Traditional Wisdom and Modern Technology in Natural Resource Management, March 13–19, 1994, Kiboswa Development Training Center, Kenya.* New York: CODEL.

Zachariah, K., and M. Vu. 1988. *World Population Projections.* Baltimore, Md.: Johns Hopkins University Press.

Index

Index

Index